Ace Academic Publishing
ACHIEVING EXCELLENCE TOGETHER

THE ONE BIG BOOK

GRADE 5

For English, Math, and Science

- Includes Math, English, Science - all in one colorful book
- Detailed instructions to teach and learn with pictures and examples
- Best book for home schooling, practicing, and teaching
- Includes answers with detailed explanations

Detailed instructions along with interesting activities

AF727416

www.aceacademicpublishing.com

Author: Ace Academic Publishing

Prepaze is a sister company of Ace Academic Publishing. Intrigued by the unending possibilities of the internet and its role in education, Prepaze was created to spread the knowledge and learning across all corners of the world through an online platform. We equip ourselves with state-of-the-art technologies so that knowledge reaches the students through the quickest and the most effective channels.

The materials for our books are written by award winning teachers with several years of teaching experience. All our books are aligned with the state standards and are widely used by many schools throughout the country.

For enquiries and bulk order, contact us at the following address:

3736, Fallon Road, #403
Dublin, CA 94568
www.aceacademicpublishing.com

Ace Academic Publishing
ACHIEVING EXCELLENCE TOGETHER

This book contains copyright material. The purchase of this material entitles the buyer to use this material for personal and classroom use only. Reproducing the content for commercial use is strictly prohibited. Contact us to learn about options to use it for an entire school district or other commercial use.

ISBN: 978-1-949383-40-9
© Ace Academic Publishing, 2019

Other books from Ace Academic Publishing

Ace Academic Publishing

ACHIEVING EXCELLENCE TOGETHER

Other books from Ace Academic Publishing

Ace Academic Publishing

ACHIEVING EXCELLENCE TOGETHER

Contents

English

Math

Science

Make Sentences

Make sentences in future tense (remember to use the right verb form.) The first one is done for you.

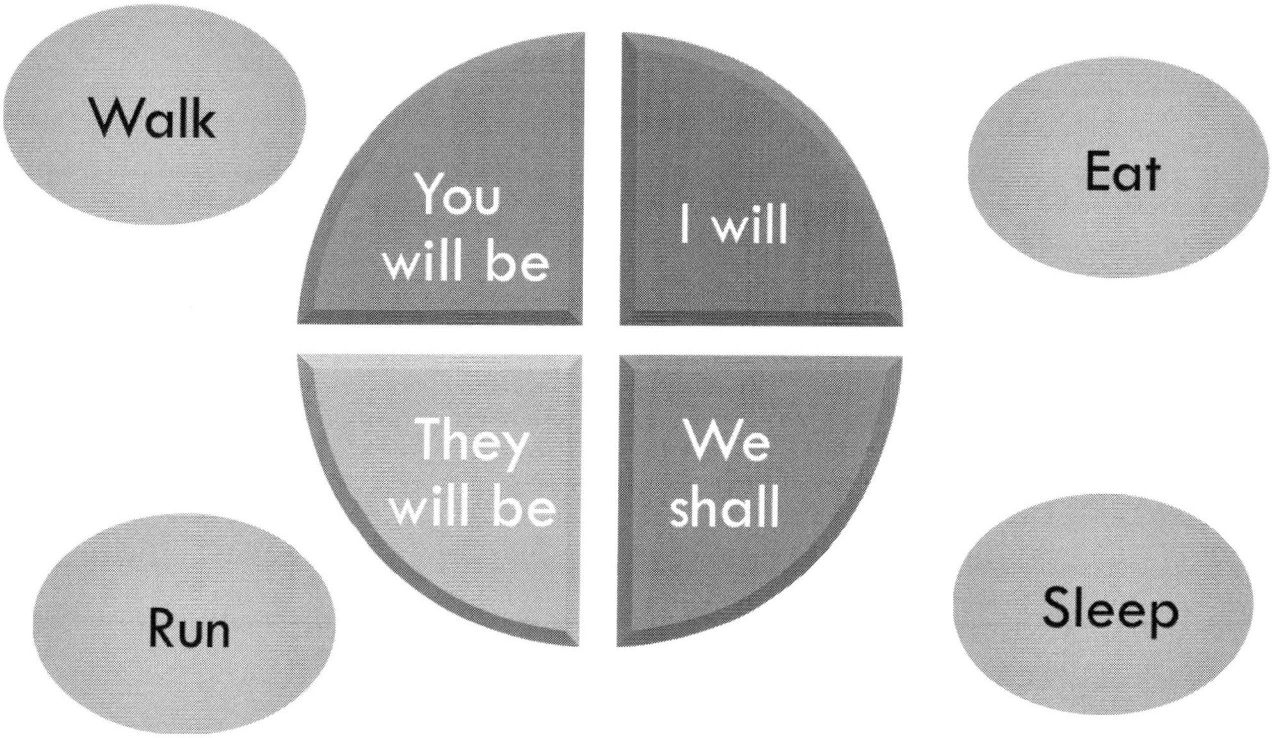

For example

Eat

1. You will be eating by the time I reach home.

2. I will eat later.

3. They will eat in the dining hall.

4. We shall be eating in a restaurant

Sleep

Run

Walk

Fill in the Blanks

Use future continuous or future simple form of the verb in brackets.

1. Will you ____________ (join) when we go to the movies tomorrow?
2. This time tomorrow, we ____________ (play) on the beach at Hawaii.
3. Don't keep the light on after midnight, I ____________ (sleep).
4. Jack says that he ____________ (submit) the assignment by Tuesday.
5. If I finish my work by afternoon, I ____________ (go) to the park by evening.

Form Future Tense

Fill in the blanks with the future perfect tense of the verbs given in parentheses.

1. When he comes next year, we ____________ to a new house. (move)
2. By the time you buy lunch, I ____________ home. (reach)
3. In five years, I ____________ my higher studies. (finish)
4. She ____________ the book before she retires. (complete)

5. My mother ______________ breakfast before you get back. (cook)

6. Please, call back later. He ______________ by two o'clock. (return)

7. I ______________ by tomorrow morning. (read)

8. Jason said that he ______________ the assignment by Monday morning. (completed)

Complete Me

Complete the sentences with the correct form of the verb given in the box.

sell	go	plan	rain	leave

1. Why don't you ask Kate? She ____________ her house in a few months from now.

2. I ____________ to Canada next year.

3. Mr. Karim, I ____________ late to school next Monday.

4. The train ____________ the station exactly at 11.45 pm.

5. See the dark clouds, I think it ____________ soon.

Conjunctions

Conjunctions are words used to connect words or parts of a sentence. There are three types of conjunctions: coordinating conjunctions, subordinating conjunctions, and correlative conjunctions.

Examples

I grew up with macaroni **and** cheese. (coordinating conjunction)

Unless you start eating healthy, you will be sick. (subordinating conjunction)

This or That?

Choose the best conjunction to fill in the blanks.

1. Dexter waited ____________ the train came. (until/yet)

2. "Hurry up, ____________ you will miss your school bus," mother exclaimed. (or, so)

3. ____________ you have a partner, you can't participate in this game. (Neither/Unless)

4. The camp leader told everyone to wake up ____________ sunrise. (before, but).

5. Sasha fell down, ____________ she didn't cry. (but, though)

6. ____________ it is 5 pm, it is still sunny. (Though, Until)

7. Choose the right word ____________ fill in the blanks. (and, because)

8. ____________ anchovies and olives are my favorite toppings. (Yet, Both)

9. ____________ Erin is clever, she is lazy. (Although, Both)

10. Sit silently, ____________ you want to see the bird. (if, and)

Combine Sentences

Combine the two sentences using the conjunction given within parentheses.

1. The boys were standing to the right of the stage. The girls were standing to the left of the stage. (and)

2. I don't have my phone. I left it in my other handbag. (because)

3. The teacher kept giving examples. The students understood the theory. (until)

4. My dog is scared of noise. I take him to the park early in the morning when it is quiet. (so)

5. I was feeling hungry. Mother asked me to wait for an hour. (but)

correlative conjunctions

We know that conjunctions are words that are used to combine parts of sentences, phrases, or clauses. Correlative conjunctions are a pair of conjunctions that connect two ideas within a sentence.

Here are a few examples of correlative conjunctions:

Either...or

Neither...nor

Both...and

Whether...or

Not only...but also

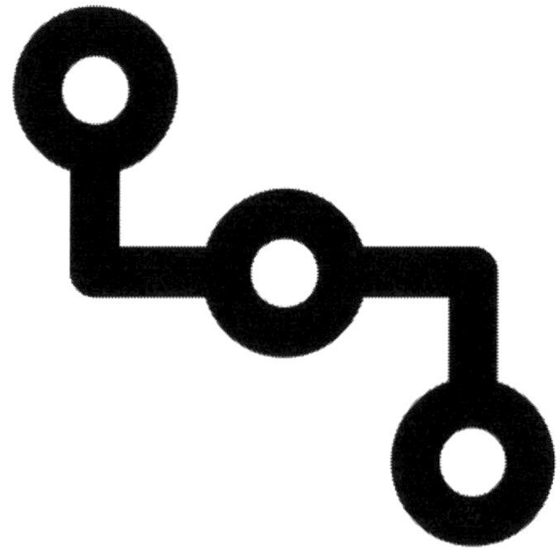

Examples

We can either go for a movie or a play.

Neither Mike nor Michael cleared the test.

Identify the conjunctions

Underline the correlative conjunctions

1. I liked both the book and the movie.
2. The doctor not only gave an injection but also prescribed a course of antibiotics.
3. You can either go shopping with your mother or stay at home and read.
4. We can neither go for swimming nor fishing today.

Complete Me!

Use the correct pair of correlative conjunctions to fill in the blanks.

1. Shirley ____________ sings beautifully, ____________ dances gracefully.
2. You can ____________ have milk ____________ cocoa.
3. This sauce is ____________ spicy, ____________ salty.
4. ____________ Peter ____________ David came to class today.
5. You can ____________ watch television ____________ play monopoly after lunch.
6. On Sundays, ____________ Eden ____________ Erin cooks dinner.

Spelling

In English language, spelling can be fun and frustrating at the same time. The vocabulary is so vast that it is hard for an average person to know all the words. Also, spelling can be challenging due to many factors such as conjugation, affixes, loanwords, homonyms, etc.

Examples

Here are a few examples of commonly misspelled words due to adding affixes or when changing the word form:

noticeable or noticable? ninety or ninty? writing or writting?

The correct spellings are in blue...

Spelling Staircase

Use the clues to find the words hidden in the staircase. (Here's your first clue – All the words hidden in the staircase start with the letter "D."

1								
2								
3								
4								
5								
6								

Clues

1. ______________ and night
2. A small writing table: ______________
3. A journal: ______________
4. You have to ______________ what you want to do this summer.
5. Blinded by bright light: ______________
6. We are going to ______________ our house for the party.

Word Game

How many words can you make from the word "SCHOLARSHIP"?

Rules

1. Don't use any letter that is not in the given word.

2. Use a letter only once in each new word you make, except for S and H, which can be used twice (as there are 2 of them).

3. Letters can be repeated in different words.

Find the Missing Letters

P		N	C		K	E	S

S	P		T		L	A

F	R		I	N	G	P	A	

	U	T		E	R

R			I	P	E

I	N	G	R		D		E	N	T

C	O		K		N	G

This or That?

Circle the word that is spelled correctly.

Calender	Calendar
Begining	Beginning
Exclucive	Exclusive
Remaining	Remeining
Portrait	Potrait
Endevour	Endeavor
Advance	Advanse
Happenning	Happening

Words within Words

Find the word that's hidden within the following words. The first one is done for you:

Father

The hidden word is "Fat"

Let's see if you can find the hidden words:

Camphor	Television	Dropper

Displayed	Followers	Through

Director

Example

Whole

Birthday

Did You Know?

The English language has more than 250,000 words, and every year new words are being added. On an average, a person knows more than 5000 words.

How many words do you know?

Context Clues

When we read, we come across many words for which we may not know the meaning. Based on the context and the surrounding words, we can understand the meaning of those difficult words.

Example

My uncle is known to be **parsimonious**. He never spends or gives any money to his daughters.

In this example, the word "parsimonious" may seem unfamiliar to many learners, but using the context "never spends or gives money," we can understand that it means "stingy."

Decode the Clues

Read the sentences and determine the contextual meanings of the underlined words based on your understanding of the sentence and prior knowledge.

1. My aunt Agatha has many <u>eccentric</u> habits. For example, she wears her Ostrich feather hat even when she's at home.

a. strange

b. normal

c. clean

2. Robert was expecting his application to be accepted. He was surprised when it was revoked.

a. acknowledged

b. forwarded

c. rejected

3. Speaking rudely to your parents is rash behavior. It will hurt your parent's feelings and prove that you don't know any manners.

a. itch

b. rude

c. funny

4. The teacher decided to award stars for every project. She thought that it would be an incentive for them to work harder on their lessons.

a. a small amount of money

b. tax

c. motivation

5. The young deer stood frozen in the glare of the headlights of the car. We turned off the lights so the deer could go back into the woods.

a. not moving

b. not visible

c. turned into ice

Affixes and Root Words

A **root word** is a word to which prefixes and suffixes are added to form a new word. A root word may or may not be a standalone word.

Prefixes	are words added to the beginning of a root or a base word to form new words.
Suffixes	are words added to the end of a root or a base word to form new words.

Examples

struct – is a root word meaning **to build.** Let's see how new words are made from it.

root word		suffix		new word
struct	+	ure	=	structure

prefix		root word		new word
in	+	struct	=	instruct

Make New Words

semi – is a prefix derived from Latin, which means partial or half.

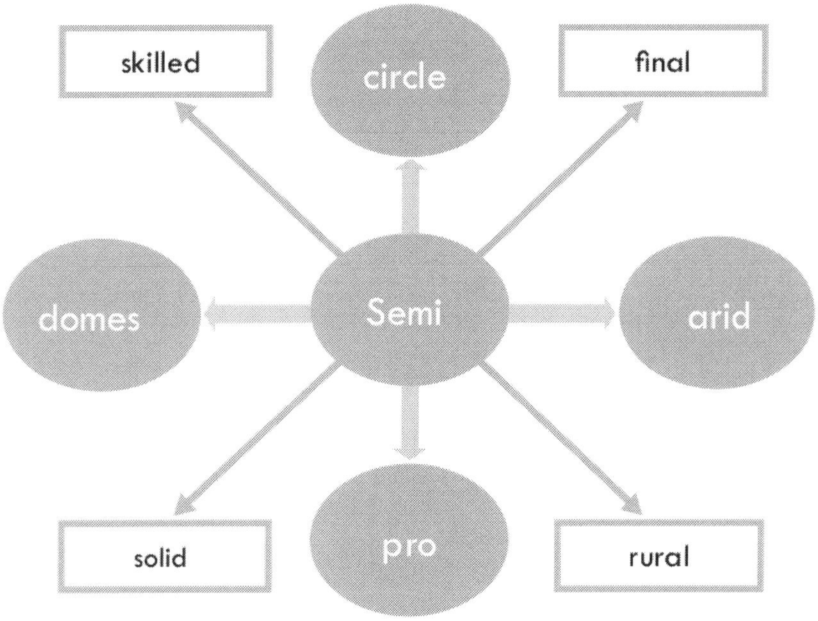

Add the prefix in the center with words around it to form new words. One is done for you.

semi + circle = semicircle

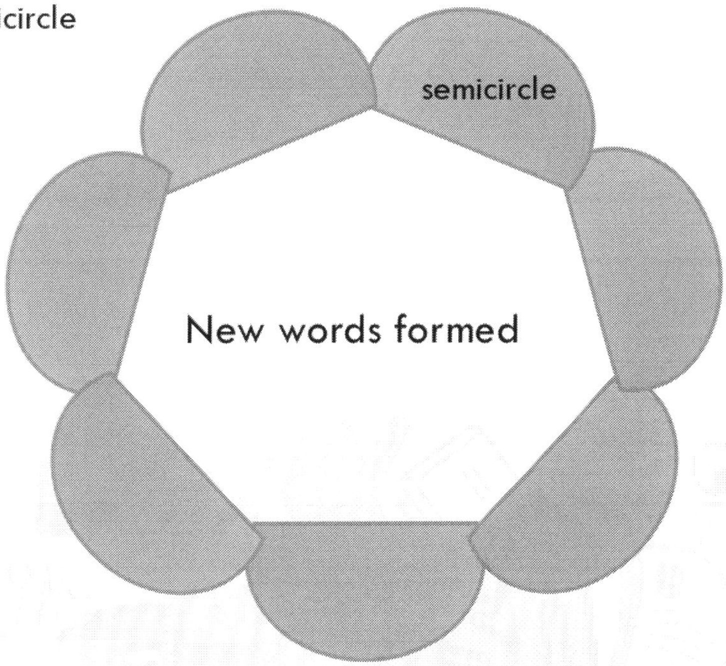

Add Suffixes

Match the suffix with its root/base words.

Root	Suffix
comfort	ary
sad	ful
diction	ly
victor	tion
hope	able
live	ship
punctuate	y
light	ness
scholar	ed
work	ing

Split the Words

Separate the root words and affixes from the list of words and write them under the appropriate column.

S.No	List of Words	Root	Prefix	Suffix
1	material			
2	judge			
3	portable			
4	adhere			
5	lovely			
6	transmit			
7	misspell			
8	subway			

Reference Materials

Have you used a Thesaurus? It is similar to a dictionary, but it helps you find synonyms for a particular word instead of meanings. Using synonyms will keep us from repeating the same word often.

Example

The synonyms of happy are

happy

joyful | delighted | cheerful | content | jovial

Find Me!

Now, use a Thesaurus in print or digital medium to find the synonyms of the following words.

story

illustrate

coat

example

more

shy

Sort Me

The box is filled with many words. Pick the synonyms of the words in the cloud and write them around the cloud.

Cure	Action	Movement	Care
Bustle	Top	Handling	Hobby
Full	Utmost	Greatest	Desk
Chart	List	Healing	Bench

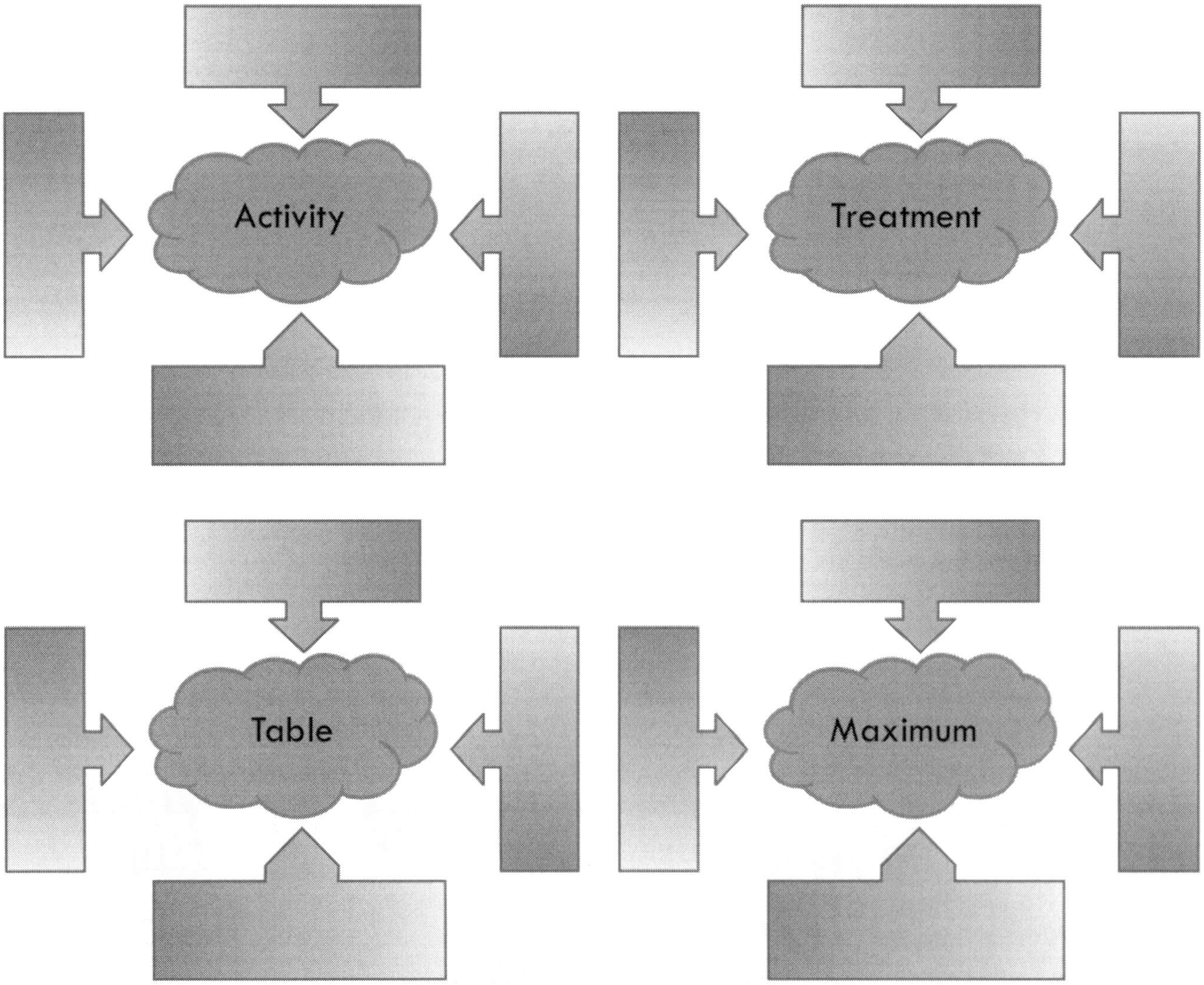

Synonyms and Antonyms

Synonyms are similar meaning words, and antonyms and opposite meaning words.

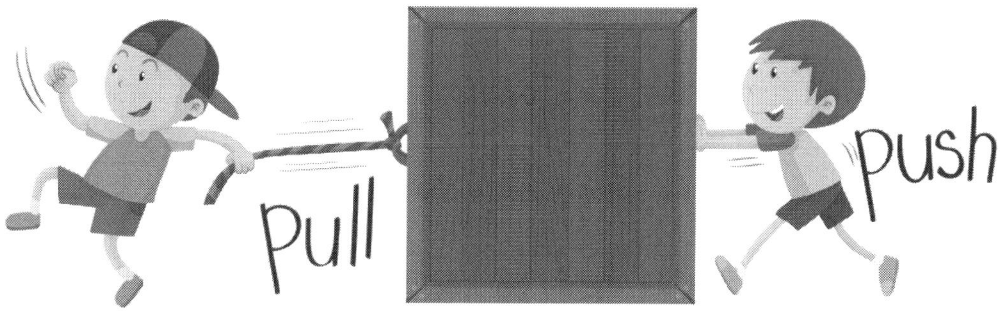

Examples

plausible - reasonable (synonym)

plausible - improbable (antonym)

Match the Synonyms

subside	utilize
benevolent	adequate
diligence	charitable
sufficient	supreme
exploit	abate
paramount	persistence

Complete Me!

Fill in the blank with the opposite of the underlined word.

1. The first question is <u>easy</u>, but the third question is ______________.

2. The hare runs ______________, but the tortoise runs <u>slow</u>.

3. Fresh vegetables are ______________. On the other hand, junk food is <u>unhealthy</u>.

4. It is difficult to ______________, but easy to <u>destroy</u>.

5. David was ______________ to climb up the tree. His sister told him to be <u>brave</u>.

Homographs

Homographs are words that have the same spelling but different meaning. These may or may not have the same pronunciation.

Example

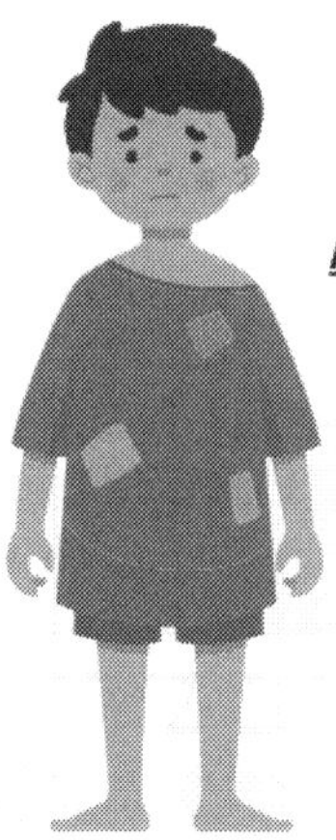

tear

<u>Meaning 1</u>: drops of water that fall from our eyes when we cry

Tears of joy fell from her eyes.

<u>Meaning 2</u>: a slit or a rip

There is a tear in your tee.

Find the Meaning

Use a dictionary to find the different meanings of the following homographs.

Bear

Minute

Close

Bank

Capital

Desert

Make Sentences

Below, we have another set of homographs. Use them in sentences to bring out the different meanings

duck

letter

novel

right

ring

Transitional Words

Transitional words or connectors are words that join two clauses. They indicate the progress of ideas between sentences and paragraphs.

Example

Sentence 1 - Samantha doesn't like burgers.

Sentence 2 - Samantha ordered burger for dinner.

We can use a contrastive transitional word to join the two sentences.

Although Samantha doesn't like burgers, she ordered it for dinner.

Connect Sentences

Use the connectors to combine the two sentences.

although	in spite of	but
however	besides	whereas

Sentence 1: We enjoyed the picnic.

Sentence 2: It rained a lot.

____

Sentence 1: I don't have a blue shirt.

Sentence 2: I have a red shirt.

____

Sentence 1: The watch was expensive.

Sentence 2: Beth bought the watch.

____

Sentence 1: It was raining.

Sentence 2: We went for a walk.

____

Sentence 1: Mother was not at home.

Sentence 2: We managed to cook lunch.

____

Sentence 1: Richard had watched the movie.

Sentence 2: Richard wanted to watch the movie again.

____

Connectors of Time

Transitional words of time indicate a relationship of time. A few examples of transitional words are: around, before, during, today, tomorrow, first, next, now, later, yesterday, when, until, etc.

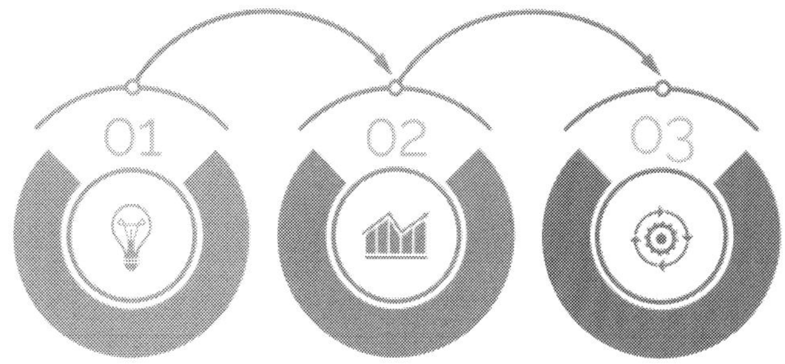

Connect the Events

Use the correct transitional words to make the below passage more meaningful.

Setting up a Vegetable Patch

Serena wanted to know how to set up a vegetable patch in the backyard. So, her mother explained it to her. She said, "(a) ______________ you should clean a small space in the backyard where you want to plant the vegetables. (b) ______________ you should plough the soil." Serena asked "Why should I plough the soil?"

Mother replied "It is important to plough the soil (c) ______________ planting. The top soil will be hard and ploughing brings the soft and loose bottom soil up. Plants grow well in such a soil."

"Ok Mom," what should I do (d) ______________ ploughing?"

"You can plant the vegetable seeds that you got." After sowing the seeds, Serena asked, "(e) ______________ will the vegetables grow?"

Mother smiled and said, "Be patient Serena. Good things take time."

Reading: Literature

Who Will Bell the Cat?

Once upon a time, in the small town of Bakerville, there was a quaint[1] bakery on Main Street. Everyone in the town would go to the bakery to buy bread, pies, muffins, bagels, bread rolls, buns, and more.

Attracted by the smell of the freshly-baked food, a family of mice came to live in the bakery. They found a hole near the rear entrance of the bakery where they decided to stay. Every day they would creep into the bakery in the middle of the night and eat all the crumbs fallen on the floor.

One day the baker had made plenty of cakes and pies. The smell was so appetizing that the mice couldn't resist themselves. They gnawed[2] through the wooden shelves and found the yummy treats. They ran on the shelves, up and down, tasting all the delicious pies, cakes, and muffins.

The next morning, the baker came into the bakery and saw all his lovely cakes in crumbles. The floor was strewn with pieces of cookies and pies. The baker got very angry and shouted, "I have to get rid of all the mice." Thinking what to do, he decided to get a cat.

The very next day, he brought a cat to the bakery. The cat ran around looking into all the corners, smelling for mice. That night, the mice came out as usual to eat the delicious pies and bread. The cat was waiting near the shelves. When the mice came out, the cat silently walked behind them and pounced on a mouse. The others seeing one mouse caught, ran away fast and hid in their hole. They were very scared and didn't come out.

The next day when the baker came into the bakery, he saw the dead mouse. He saw that all his baked goods were safe on the shelves. Feeling very happy, he treated the cat to a large fish and plenty of milk.

The cat stayed in the bakery and the mice did not come out. After a few days, the mice started getting very hungry. They tried going out to find something to eat, but every time they peeped out of the hole, they saw the cat sitting nearby looking at the hole. Scared, they would go back inside.

They called for a family meeting to decide what to do to get rid of the cat. Each mouse came up with one idea and the other, but nothing was good. Finally, one mouse said, "We are not able to hear the cat because he walks so softly. Let's tie a bell around the cat's neck, so that we can hear when he comes near."

All the mice were happy with the idea and said, "It's a great idea, let's do it."

Then an old mouse said, "Good idea, but who will bell the cat?"

[1] quaint – strange and old fashioned; [2] gnaw – to keep biting on something

True or False

1. The mice found a hole near the front entrance of the bakery.

2. The cat killed all the mice in one night.

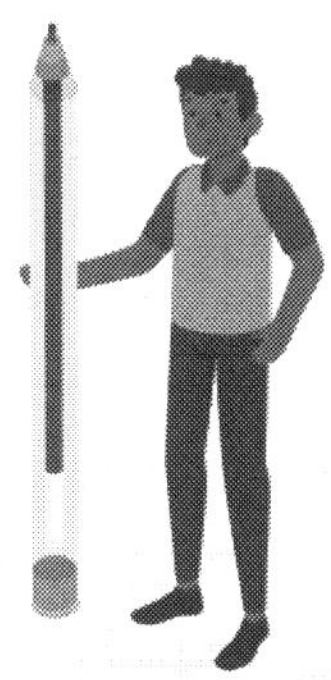

3. One night, the mice ate all the cakes and pies.

4. The cat made a lot of noise while walking.

5. The cat sat near the hole and looked at it.

Read and Answer

1. What baked goods did the baker sell?

2. How did the mice find the bakery?

3. When the baker saw all his cakes destroyed and eaten by the mice, what did he decide to do?

4. Why did the mice decide to bell the cat?

Describe My Work

A baker is a person who prepares baked goods such as bread, muffins, cakes, and more. Find out what kind of work the following people do and write your answer in the boxes.

Farmer

Cobbler

Carpenter

Butcher

Barber

Construction Worker

Find the clue

In the below section, we've got a list of words. You need to do 2 things:

1 – Find out where this word is used in the above passage and write the sentence.

2 – Circle the right meaning in the choices given below.

1. Creep

Sentence from the passage:

____

____

a. walk fast making a lot of noise

b. move slowly and carefully

c. look right and left

2. Delicious

Sentence from the passage:

____

____

a. tasty

b. salty

c. ugly

3. Crumbs

Sentence from the passage:

____

____

a. a large piece of cake

b. small broken pieces of cake

c. a slice of a cake

Write the past tense of the present tense verbs.
(Clue: The answers are in the above passage)

Present Tense	Past Tense
attract	
gnaw	
see	
give	
come	
stay	
walk	
run	
peep	
pounce	

Complete Me!

Fill in the blanks with the right verb tense.

1. When I went to the beach this morning, I ________________ a strange sight. (will see/ saw/ have seen)

2. I ________________ anything from Susan since last January. (heard/haven't heard/am not hearing)

3. "My teacher ________________ to talk to you tomorrow," Regan told his father. (wants/will want/is wanting)

4. The earth ________________ around the sun. (revolves/had revolved/will be revolving)

What Happened at the Pond?

Once upon a time there was a small boy named William. He lived with his mother and father in a small house at the edge of the town. The house was far away from the other houses in the town, and none of his friends lived nearby. During school days, William would play with his friends at school, but during the holidays, he was all alone as there was no one to play with him.

There was a small pond about five minutes away from his house. Sometimes, his father would take him to the pond for fishing on holidays or weekends.

While all the children enjoyed summer holidays, William did not enjoy it. He had to stay at home alone all day long. There was a small garden in front of his house. At a corner of the garden was a huge elm tree, where his father had hung a tire swing from a big branch. William played on the swing, and when his mother was not looking, he also tried climbing the tree. However, the branches were so high that he was not able to climb up the tree. So, William walked up and down the garden whistling and singing to himself.

One day, William was so bored that he dug up the garden to see if he could find any worms. His mother saw her beautiful garden turned into a mess and shouted "William, You naughty boy! Don't dig up all my beautiful flower beds. Go play by yourself."

William said in dismay, "But mom, there is no one to play with, and it is so boring to play all by myself all day long. Shall I go to the town and play with my friends?"

"No William, I forbid you from going to the town. It is a long walk and I will not allow you to go alone. Now, be an obedient boy and go play silently," said the mother.

Feeling very angry, William went and sat on the swing and moodily looked at the pond. Suddenly he thought, "Why not go to the pond?" He knew his mother would not allow him to go alone, so he snuck out. He ran fast all the way to the pond.

William was so happy that he started laughing and jumping. He then saw another boy standing near the pond and throwing stones into the water. William didn't know anyone living nearby. Curious, William went close to the boy. "Hi," he said. "I am William. What is your name?"

Read and Answer

1. Why was William feeling bored??

____

2. Where did William live?

____

3. Why did William go to the pond?

____

4. Who did William see at the pond?

____

5. Why was William's mom angry with him?

6. Why did William's mother not let him go play with his friends?

7. What did William do when his father had a holiday?

order Me

Arrange the events in the story in order. Write the order number in the left column.

	Feeling very bored, William digs up the flower beds
	William asks his mother if he could go to the town to visit his friends
	He then decides to go the pond
	Feeling sad, William sits on the tire swing
	William's mother scolds him for tearing up her flower beds
	William runs fast all the way to the pond
	He sees a boy standing near the pond
	It was summer holidays and William was feeling very bored
	His mother forbids him from going to the town
	William looks around silently for his mother and creeps out of the garden

Synonyms and Antonyms

Find the correct synonym and antonym of the underlined word from the choices given below:

1. William was so bored that he **dug** up the garden

a. fill

b. till

c. bill

Synonym ____________ Antonym ____________

2. William said in **dismay**, "But mom, there is no one to play with."

a. disappointment

b. surprise

c. happily

Synonym ____________ Antonym ____________

3. Now, be an **obedient** boy and go play silently.

a. busy

b. naughty

c. respectful

Synonym ____________ Antonym ____________

4. "William, You **naughty** boy!

a. obedient

b. disobedient

c. sad

Synonym ____________ Antonym ____________

5. No William, I **forbid** you from going to the town.

a. prohibit

b. allow

c. shout

Synonym ____________ Antonym ____________

Story Building

At the end of the passage, William says "Hi" to the new boy near the pond. What do you think happened next?

Write their conversation below (Give the new boy a name).

William: ....

New Boy: ....

William: ....

New Boy: ....

William: ....

New Boy: ....

William: ....

New Boy: ....

William: ....

New Boy: ....

William: ....

New Boy: ....

Vegetable Patch

Serena was a sweet girl who lived in a beautiful village that was called Cloudesberry. This village was right at the foothills of a huge mountain that was called the Cloud Mountain by the people in the village. Serena lived with her mother and father in a tiny cottage. They had a huge garden around the cottage.

This story happened more than a couple of centuries ago. So, Serena didn't have to go to school then. All day long, she played with her dog Pluto in the garden. Sometimes, her mother asked her to do some work, like fetching[1] water from the river or help with the washing. Her father was a blacksmith. Every morning, he would carry his tools and go to his shed at the center of the village for work. Mother did all the work at home. She would cook, clean, wash, sew, and also take care of the garden.

One day, Serena thought "Mother is doing so much work. I think I should help her." So, she asked her mother, "Mom, let me help you. I play all day long. Shall I do something for you?"

Mother replied, "Serena dear, don't worry. I can take care of all this. You go and have fun with Pluto and your friends."

However, Serena was determined[2] to help her mother. In her haste, Serena poured a lot of water into the flour to make the bread dough. And all the flour was wasted.

Then she tried to mop the house and ended up spilling water everywhere. Her mother almost slipped on the wet floor.

Serena wanted to help her mother with the cooking. So, she went to the stove, added salt and stirred the stew. That afternoon, no one could eat any lunch – there was no bread and the stew was too salty!

Finally, her mother thought, "If I don't get her out of the house, I will not get any chores done today." She thought what to do to keep Serena occupied and happy.

She got a fantastic[3] idea. "Serena, dear," she said, "I have a very important task for you. Can you do it?" "Surely mom," Serena replied enthusiastically. Serena's mother took her to the backyard. She said "Do you know what this place needs? A vegetable patch."

"A vegetable patch?" Serena asked, "What is it mom?"

"It's a small garden where you grow vegetables. Can you do it?"

"But mom, I don't know anything about growing vegetables," Serena replied.

"Don't worry, I will help you with everything."

"First, clean this place. Then you can grow radishes, turnips, carrots, beets, beans, and lots more. Every day we can have fresh vegetables for dinner. What do you think?"

"I think it is a fantastic idea mom," she said. "Pluto and I are going to work hard on this vegetable patch. Ok, you go on, I have a lot of work to do. Don't disturb me."

Serena's mom smiled at her lovely daughter and went back to complete her chores in peace.

[1] fetching -bring; [2] determined – firm and strong minded; [3] fantastic – great, unbelievable

Story Elements

1. This village was right at the foothills of a huge mountain called ______________.

2. Serena's father was a ______________.

3. Pluto was Serena's ______________.

4. You can grow ______________, ______________, and ______________ in the vegetable patch.

5. ______________ did all the work at home.

Crossword Puzzle

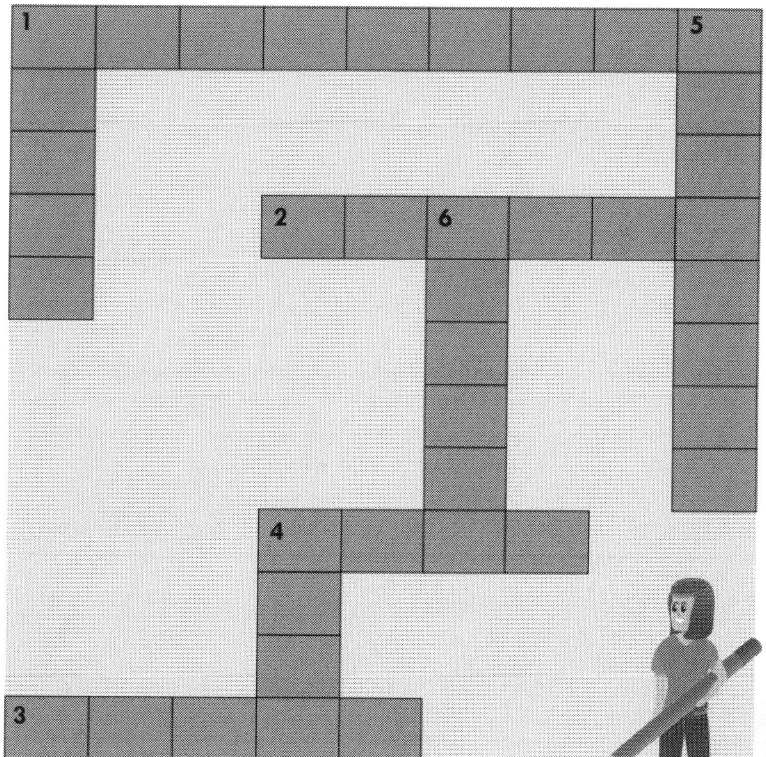

Across

1. It's both a country and a continent (9)
2. I've had ______________ (6) (clue-starts with "E")
3. Broken pieces of cake (6)
4. An adult plant (4)

Down

1. A word that means to do once more (5)
4. A group that works together (4)
5. It's a subordinate conjunction (8) (clue-starts with "A")
6. A color and a fruit with the same name (6)

Adjectives

In the passage, adjectives have been used to say more about the person or the place. For example:

Sweet girl – Here the word "sweet" describes the girl was kind and charming

Beautiful village – Here the word "beautiful" describes the village

Use the right adjectives from the below list to describe the person or place.

Beautiful	Tall	Dark
Dirty	Fresh	Honest
Green	Huge	

Tea

Room

Vegetable

Flower

Mountain

Building

Water

Worker

Character Analysis

We can say a lot about a person by their actions. In this story, Serena does a lot of things. What do the actions say about Serena's character?

1. Sometimes, her mother asked her to do some work, like fetching water from the river or help with the washing.

The above lines mean that Serena helped her mother with the chores. It shows that she was

Lazy | Bored | Obedient

2. "Pluto and I are going to work hard on this vegetable patch." - These words show that Serena was

Hard working | Beautiful | Sloppy

3. In her haste, Serena poured a lot of water into the flour to make the bread dough. - This action shows that Serena was

Impulsive | Thoughtful | Careful

4. Mom, let me help you. I play all day long. Shall I do something for you?" - These words show that Serene was

Selfish | Helpful | Careless

5. However, Serena was determined to help her mother. - It shows that Serena was

Strong Minded | Weak Minded | Brilliant

The Song of the Bee

Buzz! buzz! buzz!
This is the song of the bee.
His legs are of yellow;
A jolly, good fellow,
And yet a great worker is he.

In days that are sunny
He's getting his honey;
In days that are cloudy
He's making his wax:
On pinks and on lilies,
And gay daffodillies,
And columbine blossoms,
He levies a tax!

Buzz! buzz! buzz!
The sweet-smelling clover,
He, humming, hangs over;
The scent of the roses
Makes fragrant his wings:
He never gets lazy;
From thistle and daisy,
And weeds of the meadow,
Some treasure he brings.

Buzz! buzz! buzz!
From morning's first light
Till the coming of night,
He's singing and toiling
The summer day through.
Oh! we may get weary,

And think work is dreary;
'Tis harder by far
To have nothing to do.

by Marian Douglas

Poem Appreciation

1. Who is a jolly good fellow?

2. What does the bee do on sunny days?

3. What does the bee do on cloudy days?

4. From where does the bee get honey?

5. What is harder than working all day?

Character Analysis

What can be said about the character of the bee from the poem?
Choose the right answers and make a sentence using the words.

Hard Working	Lazy	Happy
Tired	Sad	Energetic

Words that rhyme

Words that have the same or similar ending sounds are called rhyming words. Examples:

see	bee
bat	hat

Rhyming Words

Find words from the poem that rhyme with the below words.

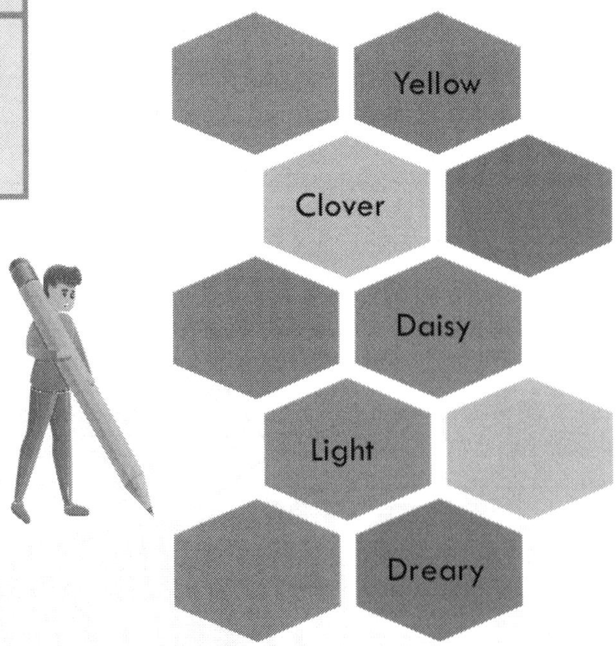

Paint the Sentence!

In this exercise, you are given 2 sentences. You need to bring life to these sentences with your colorful imagination. Write a short narrative of what may have happened.

You need to use the given sentence at least once in the narrative. So, let's begin!

1. "Mother," she shouted, "Come and see what I found!"

2. Guess where we are going this summer?

The Story of How Payton Made Friends!

Do you know the story of how Payton made friends?

Wait! Do you even know who Payton is?

I think that the best place to start a story is at the beginning. Well, here's the story of this lonely little boy called Payton. He was ten years old. He was taller than all his classmates and wore glasses.

Payton didn't like the fact that he was the tallest in his class, so he slouched[1] all the time. He thought that because he was so tall, no one wanted to be friends with him, but the truth was that all his classmates were in awe[2] of his height.

Payton's mother would tell him, "Don't hunch over, Payton, walk straight and tall." He would reply, "Yes Mother," but then would slouch again. Even when he went to school, he would walk with his head down holding his books close to him. He would go to the last seat in the classroom and sit. Payton would only talk if the teacher asked him a question. His classmates tried talking to him, but he was too conscious about himself to talk. This went on for months and soon the others in the class stopped trying to talk to him.

Then one day, Ms Alan, their teacher, announced try-outs for junior basketball team. It was a pandemonium[3] during the try-outs. Payton was sitting silently in the corner. The teacher looked at Payton and said, "Payton, why don't you try?" Payton hesitated thinking, "What if I humiliate myself and everyone laughs at me?" With Ms Alan's encouragement, he mustered courage and tried out. He made a basket, and everyone cheered for him.

Payton got selected for the team and felt very happy. The others came to him and shook his hands. He replied with enthusiasm and felt confident to make friends. It was a new beginning for him.

[1] slouch: hunch over; [2] awe: a feeling of respect with wonder or fear;
[3] pandemonium: confusion or wild and noisy disorder

Story Elements

1. Why did Payton hunch over all the time?

___

2. How old is Payton?

___

3. Did Payton have many friends? Why?

___

4. What try-out was announced in class?

___

5. Why do you think Payton decided to go for the basketball team selections?

___

Connecting to Text

Have you experienced anything similar to Payton?

Have you gone out of your way to help someone make friends?

How can someone make a boy like Payton feel comfortable in the class?

Jumbled Sentences

Rearrange the words to make meaningful sentences.

would He seat go to last sit in classroom the and the

Well, the story here's called of little boy Payton lonely this

Payton was like his fact the that he the in class didn't tallest

and The said. looked at teacher "Payton, why try you?" Payton don't

Similar Words

Find a word in the passage that means the same as the following words:

Start	
Truth	
Choice	
Humiliate	

Unscramble the Words

H	E	G	U	N	O	
I	S	E	L	P	M	
D	W	L	R	S	O	
O	R	R	C	D	E	
A	C	R	E	L	O	
I	I	T	N	E	V	

A Good Play

We built a ship upon the stairs
All made of the back-bedroom chairs,
And filled it full of sofa pillows
To go a-sailing on the billows.

We took a saw and several nails,
And water in the nursery pails;
And Tom said, "Let us also take
An apple and a slice of cake;"—
Which was enough for Tom and me
To go a-sailing on, till tea.

We sailed along for days and days,
And had the very best of plays;
But Tom fell out and hurt his knee,
So there was no one left but me.

by R L Stevenson

Poem Appreciation

1. Where did they build the ship?

___

2. Why did they build a ship?

___

3. With what did they build the ship?

4. *Which was enough for Tom and me*
Who is referred to as "me"?

5. Who is the narrator?

6. Why was the narrator all alone at the end?

Connecting to Text

How is the author and Tom happy playing with imaginary ship made out of furniture?

Do you find happiness with things you can find around the house or do you need expensive things to play? Why?

Why should we find happiness in simple things?

Rhyme Scheme

Rhymes are words that have a similar ending sound. Poets use these rhyming words in different orders in their poems.

For example, in the first stanza of this poem, the first two lines end with **stairs** and **chairs**, which are rhyming words.

The third and the fourth lines end with **pillows** and **billows**, which rhyme with each other.

This rhyming scheme is AABB

Write an 8-line poem with AABB rhyme scheme.

Word Connect

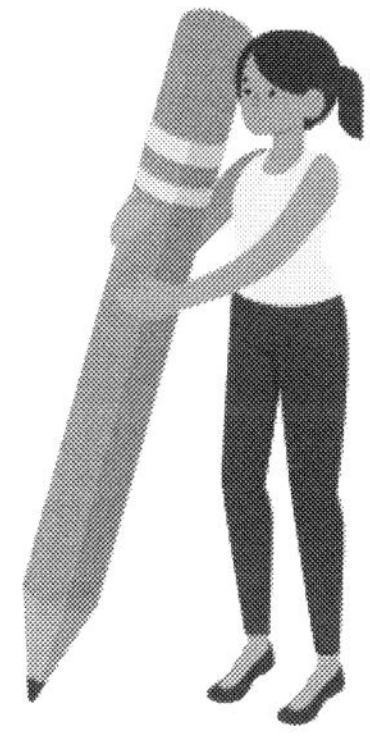

Connect the letters to find the hidden words. The first one is done for you.

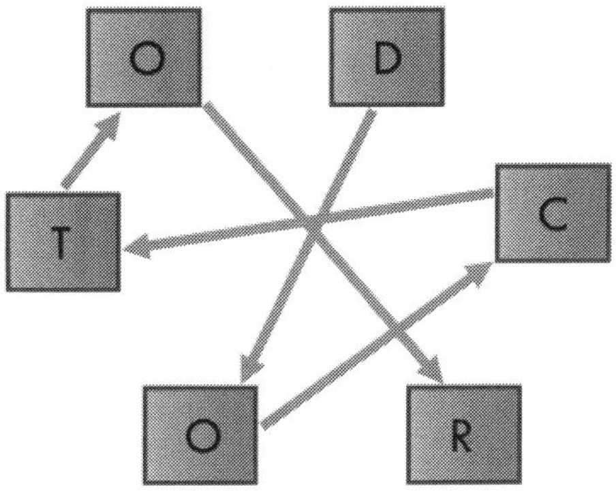

DOCTOR

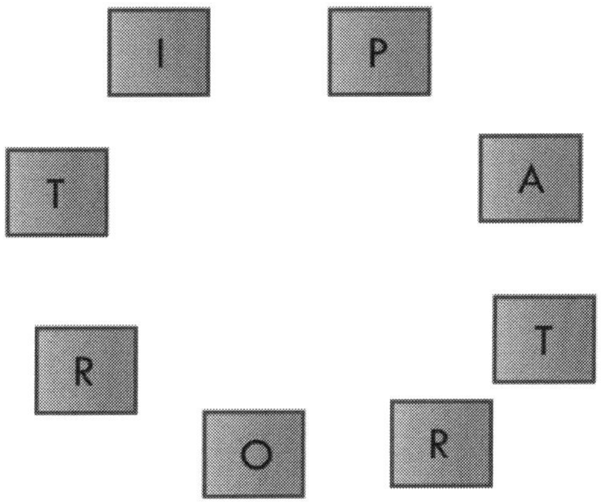

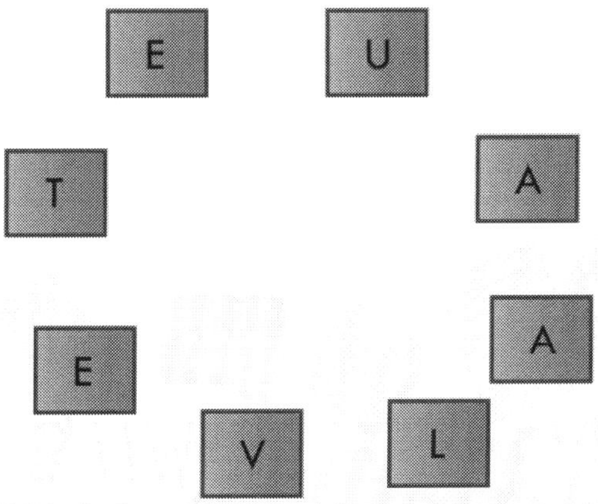

P T A H C

P E A L M S

Did You Know?

A pangram is a sentence in which every letter is used at least once.

Example

Watch "Jeopardy!" Alex Trebek's fun TV quiz game.

Can you try writing a pangram?

Reading: Informational Text

Lion and Tiger

You may have seen these two large animals in a zoo or maybe in a movie! Or do you remember reading about these wild animals? Though the two animals live in the depth of the forests, they are different from each other in many ways.

Lion

- Lions belong to the cat family.
- They mostly live in the grasslands and sometimes forests.
- They are carivorous.
- They are social animals.
- Female lions hunt together.
- In wild, lions are found in Africa and India.
- Body is tan colored.
- Average lifespan in wild is 12 years.
- Average lifespan in captivity is 15 to 16 years.
- Bite pressure is more that 1000 psi[1].

Tiger

- Belong to the cat family.
- They mostly live in the forests.
- They are carnivorous.
- They are solitary animals.
- Tigers mostly hunt alone.
- In wild, tigers are found in Southeast and Eastern Asia.
- Have stripes on the body.
- Average lifespan in wild is 15 to 20 years.
- Average lifespan in captivity is 20 to 26 years.
- Bite pressure is more that 1000 psi.

[1] psi: per square inch

Compare Texts

1. In what ways are the lion and the tiger similar?

____

____

2. In what ways are the lion and the tiger different (any two)?

____

____

3. Where can you see the lion in wild?

____

4. Where can you see the tiger in wild?

____

5. What do you think bite pressure means?

____

Stephen Hawking

1. Stephen Hawking was born on January 8 in the year 1942.

2. Hawking won a scholarship in Natural Sciences in the year 1959 when he was 17 years old.

3. He graduated college in the year 1961.

4. In the year 1962, he was diagnosed with a nerve disease.

5. From 1965 to 1970, Hawking worked on the Theory of General Relativity.

6. Stephen Hawking completed his doctorate in 1966.

7. Stephen Hawking was awarded the Eddington Medal in 1975.

8. Stephen Hawking received the Hughes Medal in 1976.

9. In 1977, Hawking became the professor of Gravitational Physics at Cambridge.

10. In 1988, Stephen Hawking's *Brief History of Time* was published.

11. Stephen Hawking published the book *On the Shoulder of Giants* in 2002.

12. He received the Michelson Morley Award of Case Western Reserve University in the year 2003.

13. Hawking was given the Copley Medal in 2006.

14. In April 2007, Hawking went on a zero-gravity flight at the Kenney Space Center.

15. Stephen Hawking died on March 14, 2018.

Read the points on Stephen Hawking and answer the following questions.

1. Which year was Stephen Hawking born?

a) 1938

b) 1942

c) 1959

2. Hawking won a scholarship in ______________ when he was 17 years old.

a) space center

b) gravitational physics

c) natural sciences

3. Which of these books were published in the year 2002?

a) A *Brief History of Time*

b) *On the Shoulder of Giants*

c) *My Brief History*

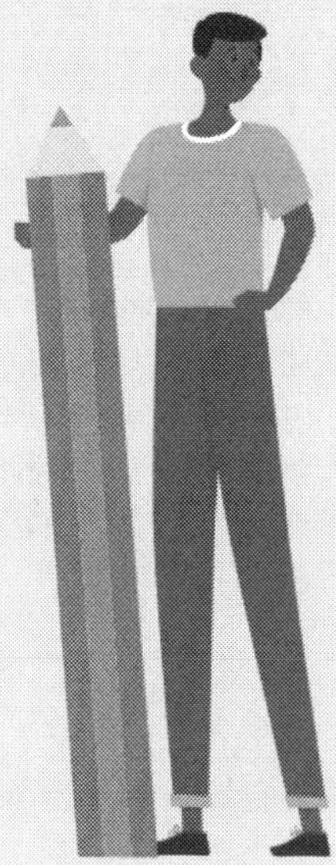

4. What happened in 1977?

a) completed his doctorate

b) published his first book

c) became a professor at Cambridge

5. Which of these medals were awarded to Stephen Hawking?

a) Eddington Medal

b) Copley Medal

c) Both a and b

Describing Words

Identify the expressions.

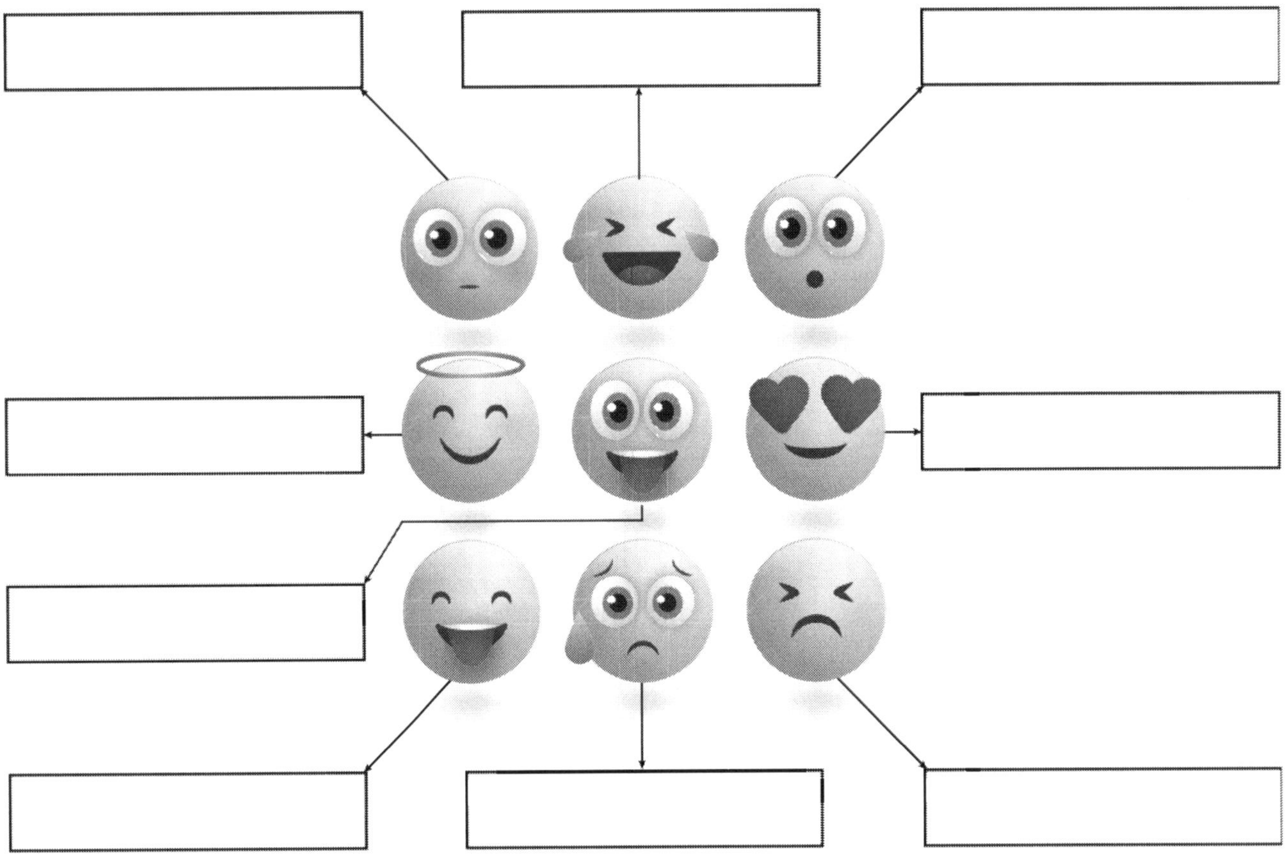

Complete Me!

Fill in the blanks with the correct form of the word given in parentheses.

1. I love these shoes as they are very ____________. (comfort)

2. Eating fried food is not very ____________. (health)

3. Margi likes wearing ____________ clothes. (fashion)

4. My mother enjoys watching ____________ movies. (class)

5. When you have a ____________ smile, you will have crinkles around the eyes. (genuinely)

6. A happy smile will get you a ____________ reaction from people. (positivity)

7. Douglas likes to wear ____________ shirts. (color)

8. Body language is one of the forms of ____________ communication. (non-verbally)

Plant Growth

The table below lists the growth of three different plants in low light, partial light, and full light.

Plant	Amount of Light	Day 5	Day 10	Day 15	Day 20	Day 25	Day 30
Plant A	Full Light	1 cm	2 cm	4 cm	8 cm	16 cm	22 cm
Plant B	Partial Light	1 cm	5 cm	8 cm	15 cm	25 cm	32 cm
Plant C	Low Light	1 cm	3 cm	6.5 cm	12 cm	21 cm	26 cm

* cm refers to centimeter

Interpret Data

1. For how many days was the experiment conducted?

____

2. Based on the observations, which kind of light is best for the plant growth?

____

3. What is the height of the plant grown under partial light on Day 15?

---

4. What was the height of the plant grown under full light on Day 10?

---

5. What is the maximum height that the plant can grow in a month and under which kind of light?

Connecting to Text

1. What kind of plants do you have in your house/school/neighborhood?

---

2. Do those plants get enough sunlight?

---

3. What else do plants need to grow besides light?

Find the Hidden objects

Identify the objects in the room and list them below.

Writing

Narrative Writing

Write a story of your own based on the below pictures.

Before you start to write the story, below is an overview of the **writing process.**

Research
Gather all the details you need to write the story. In this case, these pictures will help you write the story.

First Draft
Organize all the facts in a neat manner. Arrange the events of the story in order of occurrence. Use lists or bullet points.

Revision
Go through all that you have written to check if you have all the information right.

Final Draft
Prepare another draft with all the corrections. Your Story is Ready!

Instruction

- Refer to all the pictures in your story.
- You can refer to one picture as many times as you want.
- The pictures given below are in no particular order. So, you can arrange the events in any order.
- Add other characters and make the story interesting!

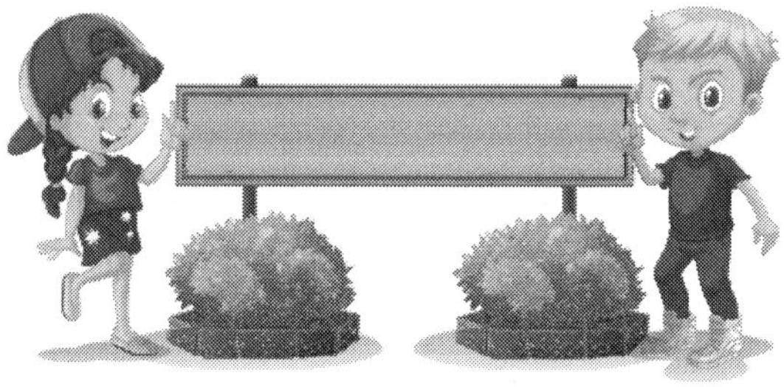

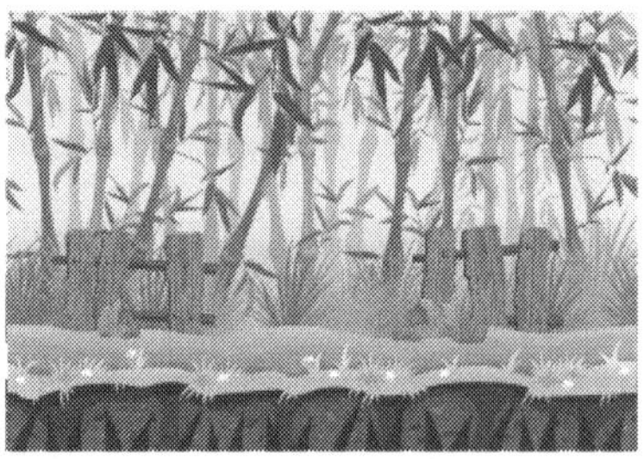

First Draft

Revision

Congrats! You have completed the first draft. Now, you need to check what you have written. You can also ask an adult to read your first draft and give you suggestions to write it better.

Checklist

- [ ] Include all the pictures
- [ ] Give a catchy title
- [ ] Don't forget to name the characters in your story
- [ ] Organize all the events in the story
- [ ] Give a conclusion

Final Draft

Opinion Writing

Write an essay comparing two wild animals. Give your opinion stating which animal you prefer and give reasons for your stand.

Do some research to write about the two animals.

Tips to organize your essay

- Give an introduction about the two animals including your opinion.
- Talk about the similarities.
- Describe the differences.
- Conclude restating your opinion.

Review Writing

Reviews are written by customers who buy things from a store. The review will say what the customer liked or didn't like about the product. Reviews are more like opinions about the product. For example

"This ice cream tastes awesome. I love the combination of chocolate and nuts." I will definitely recommend this to my friends and family.

Now, imagine you have bought a few things, watched a new movie/show, or read a book. How would you review them? You can also ask your parents to comment about things they bought recently, and write based on it!

Things	Review
Shirt	This shirt is awful! The first time I washed it, it shrunk. No, it doesn't fit me. Also, the color is fading. It was a waste of money.

Informational writing

Write a summary of the experiment based on the notes given below:

- 3 plants were taken for the experiment – Plant A, Plant B, Plant C
- Plant A - under full light
- Plant B - under partial light
- Plant C - under low light
- All plants were 1 cm high at the start of the experiment
- Experiment conducted for one month
- Result – Plant B grown under partial light showed best growth (32 cm)

Research Activity

Research on Stephen Hawking with the help of your parents or books or digital media. Put together a paragraph describing his life and work.

Journal Writing

Journals are a great way to preserve the beautiful memories. You can also write about your dreams and aspirations, happy moments, sad times, plans, and anything that you like.

Rules for Journal Writing

Remember to write the date

Mention all the details like where you were, who you were with, what you were doing

Write about your feelings

If you're making a plan, write about when you want to start doing it

Write a journal entry for today.

Math

Use this book to enable your children to explore numbers by solving interesting puzzles and real-life problems. Engage your children with fun, colorful activities and let them fall in love with Math.

Numerical Expressions

A numerical expression is a combination of numbers, integers combined using mathematical operators such as addition, subtraction, multiplication, or division.

An expression can also be a combination of variables and constants using mathematical operations. Such an expression is known as an algebraic expression.

Examples

a. $11 \times (11 - 1) + (10 \div 2)$

b. $2 + j + (8j - 1)$

Let's practice

Calculator Fun

Observe the calculator. The equation is given in the display of the calculator and the answer is circled in green. Draw parentheses to make the equations true. One is done for you.

$18 \div (9 - 6) = 6$

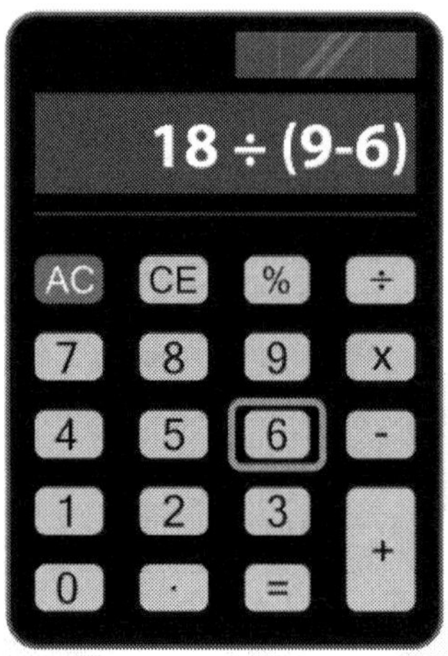

a.

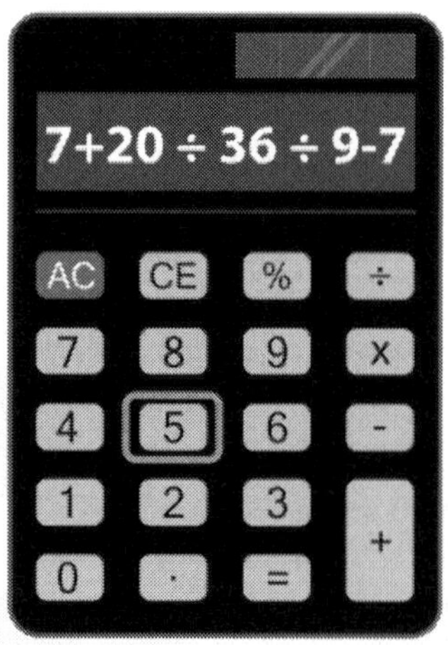

b.

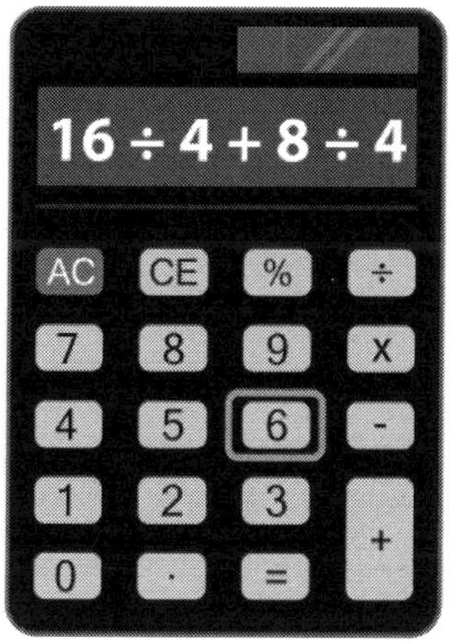

c.

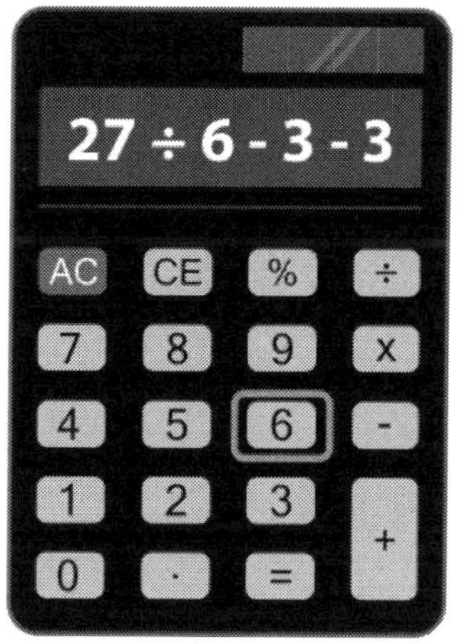

d.

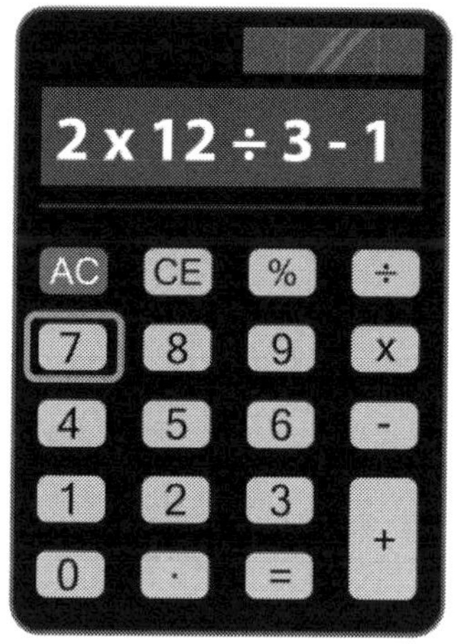

Which Car Will Reach the Gas Station?

Write a numerical expression for the statements in the table. Color the route of the car that reaches the gas station and write its expression in the yellow box.

Car	Statement	Expression
A	Add $\frac{5}{4}$ and $\frac{7}{4}$ then multiply by 2	
B	Subtract $\frac{9}{2}$ from $\frac{15}{2}$ then divide by $\frac{5}{12}$	
C	Sum of 8 tenths and $\frac{5}{8}$ is multiplied by $\frac{2}{5}$	
D	Multiply 9 by 9 then subtract the product from 98	
E	Divide $\frac{7}{8}$ by the difference between 12 and 5	

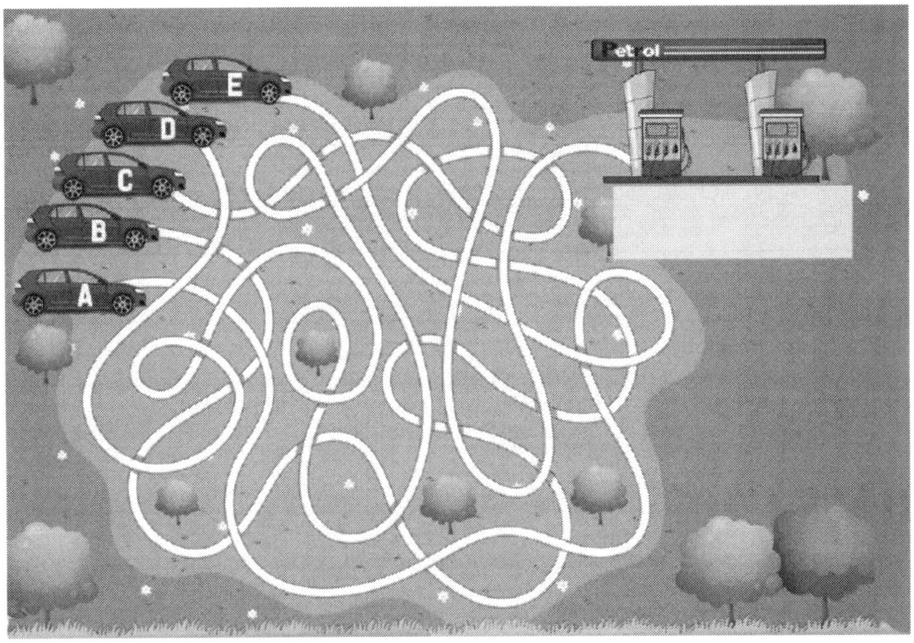

Simplify the Expressions

Simplify the following expressions. One is done for you.

(3 x 12) + 3 ÷ 3 36 + 3 ÷ 3 36 + 1 37	a. 6 x (11 – 1) + (14 ÷ 2)	b. 50 – 2 x (3 + 3)

c. $(\frac{2}{5} \times \frac{5}{6}) \div 9$	d. $0.7 - 5 \times 0.4 + (0.6 \times 6)$	e. $(1\frac{1}{4} - \frac{1}{2}) \div (1 - \frac{1}{3})$

Arrange the Markers

Solve each expressions based on the answers arrange them in ascending order from A to H in the table given below.

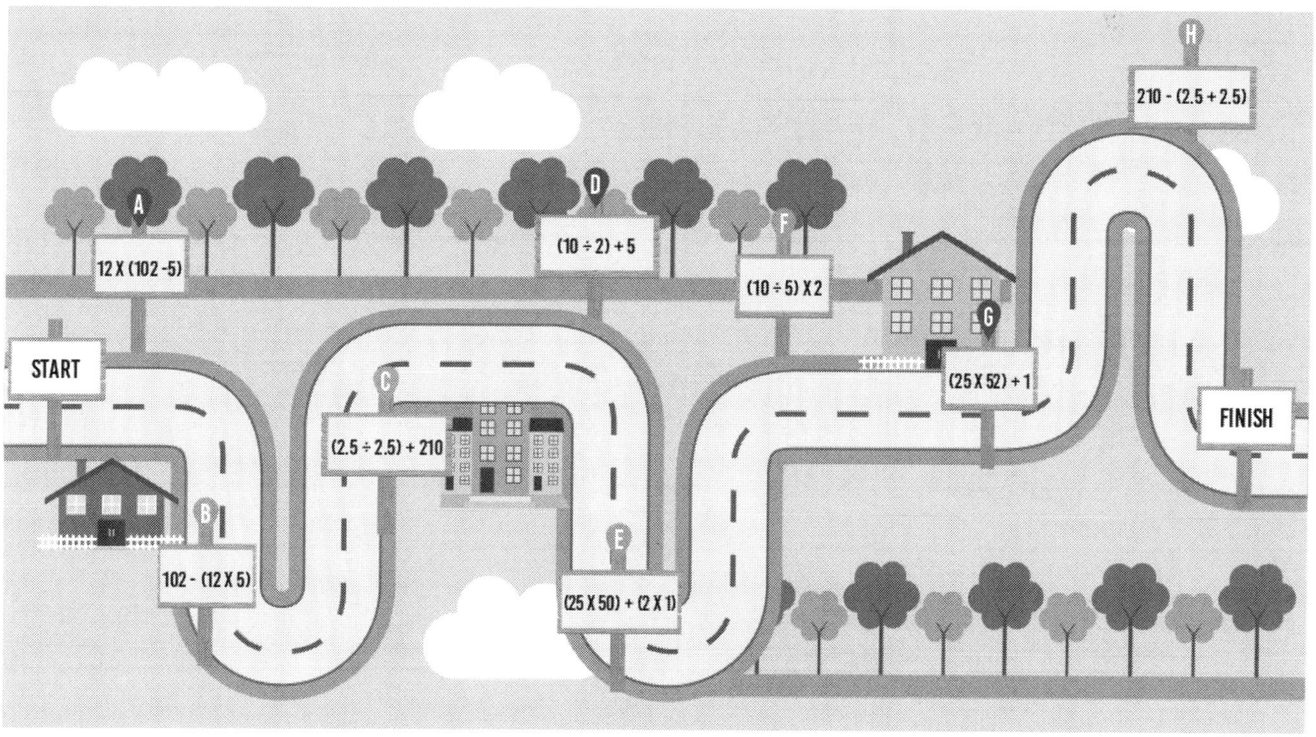

Marker A		Marker E	
Marker B		Marker F	
Marker C		Marker G	
Marker D		Marker H	

Story Problems

Read the story problems. Choose an expression below that matches the story problem, and write it in the blank.

a. Mr. Charles picked 10 pumpkins from his farm. He cooked $\frac{1}{4}$ of them, and then gave 5 to his relatives. Write the expression that tells how many pumpkins he had left.

Expression:

b. Mr. Charles picked 10 pumpkins from his farm. He cooked 5 of them, and then gave $\frac{1}{4}$ of the remaining pumpkins to his relatives. Write the expression that tells how many pumpkins he had left.

Expression:

What is the cost of Dinner?

The following expression shows the cost of dinner(in dollars) for n people.

2 + 5n + (2n - 1) x 2

Use the above expression to find the:

Cost for 1 person:

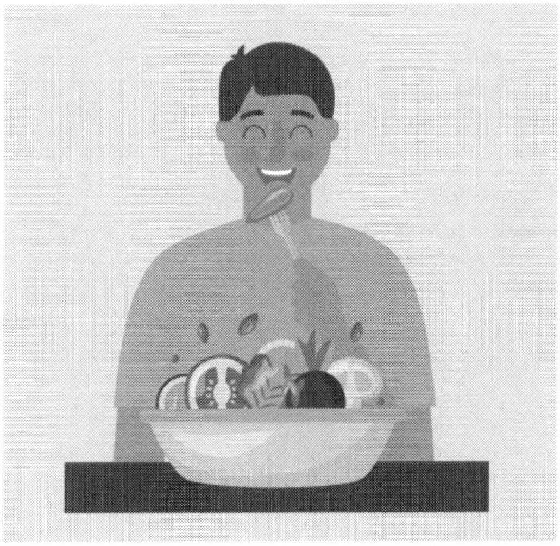

Cost for 5 people:

Fill the Jars!

Compare the numerical expressions without calculating. Write the greater expression in the empty jar. If both the expressions are equal write them on the sticky note.

a. $134 \times (11 + \frac{7}{8})$; $(134 \times 11) + \frac{7}{8}$

b. $(2.1 + 1.45) \times \frac{3}{4}$; $\frac{3}{4} \times (2.1 + 1.45)$

c. $\frac{1}{4} + (\frac{1}{2} \times 1.4)$; $(\frac{1}{2} + \frac{1}{4}) \times 1.4$

d. $95 \times (62 + \frac{7}{9})$; $(95 \times 62) + \frac{7}{9}$

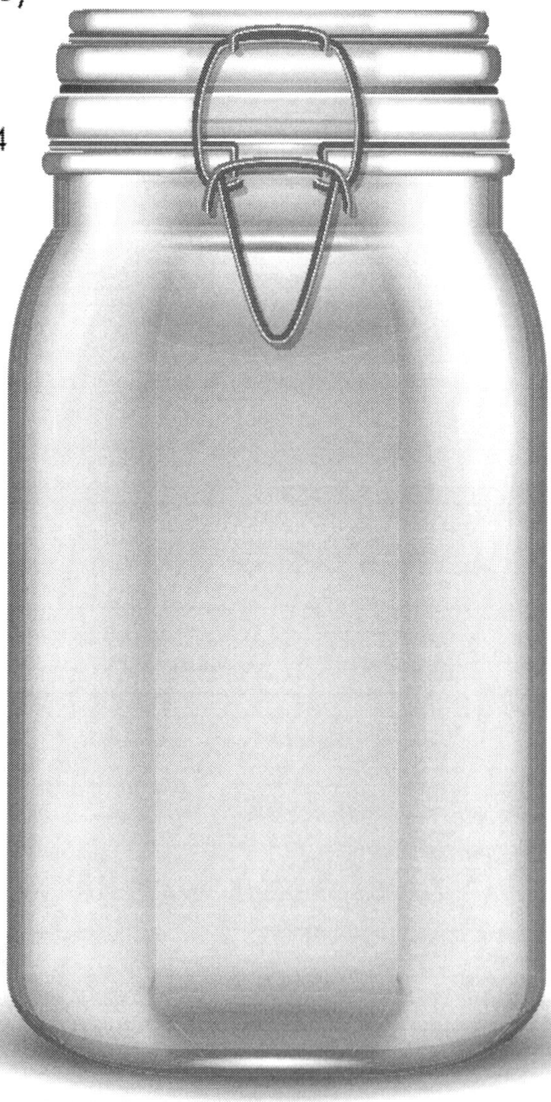

Match each numerical expression with its word form. Color each pair in the same color. One is done for you.

$12 \div (4 + 2)$	The quotient of 10 and 5 times $\frac{1}{4}$	$(6 + 8) \div \frac{1}{5}$	$(\frac{1}{4} + \frac{3}{4}) \div \frac{7}{4}$
The sum of $\frac{1}{4}$ and $\frac{3}{4}$ divided by $\frac{7}{4}$	Triple the sum of 3.5 and $\frac{9}{8}$	Twelve divided by the sum of 4 and 2	The sum of 6 and 8 divided by $\frac{1}{5}$
$\frac{1}{2} \div (\frac{5}{2} - \frac{1}{4})$	$(10 \div 5) \times \frac{1}{4}$	$(3.5 + \frac{9}{8}) \times 3$	Half divided by the difference between $\frac{5}{2}$ and $\frac{1}{4}$

Choose the correct option

Choose the option that is **not equivalent** to the expression or answer given.

a. 16×29

A 16 x (30 - 1)

B 29 times sixteens

C (10 x 29) - (6 x 29)

D (15 + 1) x 29

b. 552

A 8 x 7 x (6 + 4) - 8

B 560 - 8

C (10 x 8) - (8 x 7)

D (8 x 7) x 10 - 8

c. 120 + 95

A (6 x 2) x 10 + 100 - 5

B 12 x 10 + (100 + 5)

C 12 x (5 x 2) + (85 + 10)

D (12 x 10) + (90 + 5)

d. 325

A (600 x 2) + (30 - 5)

B (600 ÷ 2) + (5 x 5)

C 50 x (3 x 2) + 25

D (60 x 5) + (15 + 10)

Richard delivers 180 newspapers every day. Felix delivers 135 more newspapers each day than Richard.

a. Write a numerical expression to show how many newspapers Felix will deliver in 30 days.

b. Use mental math to solve the expression you had written. Write the answer here.

Number Operations

A decimal number is a number that uses place value and a decimal point to show values less than one, such as tenths and hundredths.

A fraction represents equal parts of a whole or a collection.

Let's practice

Match the following.

1/10 of 3	0.03
10 times greater than 300	3
1/10 of 3000	0.3
10 times more than 0.30	30
1/100 of 3	300
1/10 of 300	3000

Look at the two numbers on the wooden board and answer the questions.

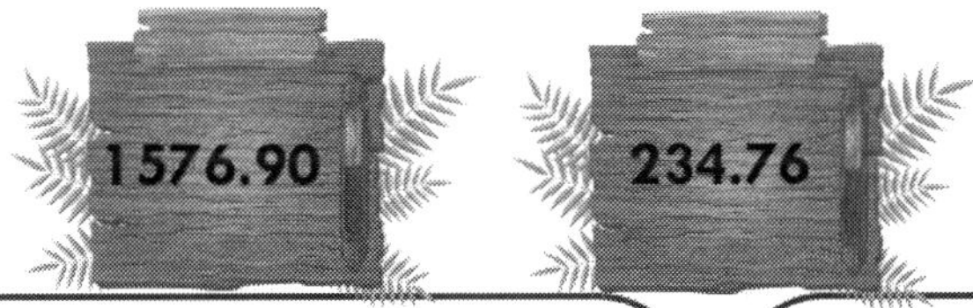

a. Which of these two has the larger place value of 7?

b. Write a statement comparing the place values of 7 in both the numbers?

c. Explain your reasoning.

Word Problems

a. If 4 pounds of flour cost $15.40, how much will 40 pounds of flour cost?

b. A dollar bill measures 6 inches. How many inches will 10 one-dollar bills measure?

c. If 2 sweaters cost $156.78, what will be the cost of 200 sweaters?

Compare

Compare the following using the symbols <, >, and =.

a.

b.

3 THOUSANDTHS + 4 HUNDREDTHS ◯ 0.034

c.

4 X 1000 +
3 X 100 +
2 X 1/100 ◯ 4300.02

d.

6 X 1/10 +
4 X 1/1000 ◯ 0.640

e.

Write the word form and expanded form.

a. 345.678

Word Form

Expanded Form:

b. 112.345

Word Form

Expanded Form:

c. 124.009

Word Form

Expanded Form:

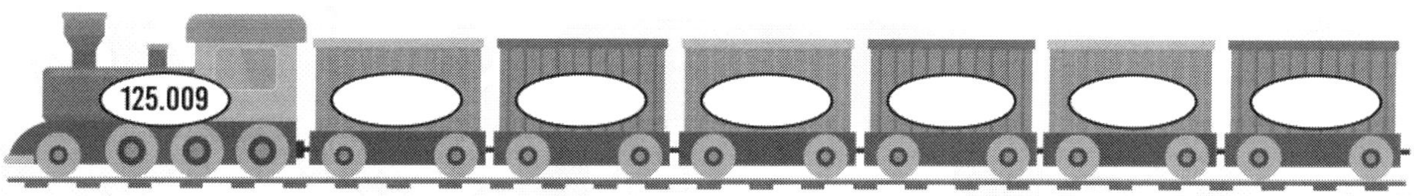

Solve the following equations by using the place value chart and arrows to show how the value of the digits in the number change.

a. 16.87 x 100

thousands	hundreds	tens	ones	.	tenths	hundredths
		1	6	.	8	7

b. 6.87 x 10

thousands	hundreds	tens	ones	.	tenths	hundredths
			6	.	8	7

c. 687 ÷ 10

thousands	hundreds	tens	ones	.	tenths	hundredths
	6	8	7			

Ava is an urban planner reviewing a site plan.The area of the neighborhood in the plan is 0.451 square meter. The actual area of the neighborhood is 1000 times the representation on the plan.

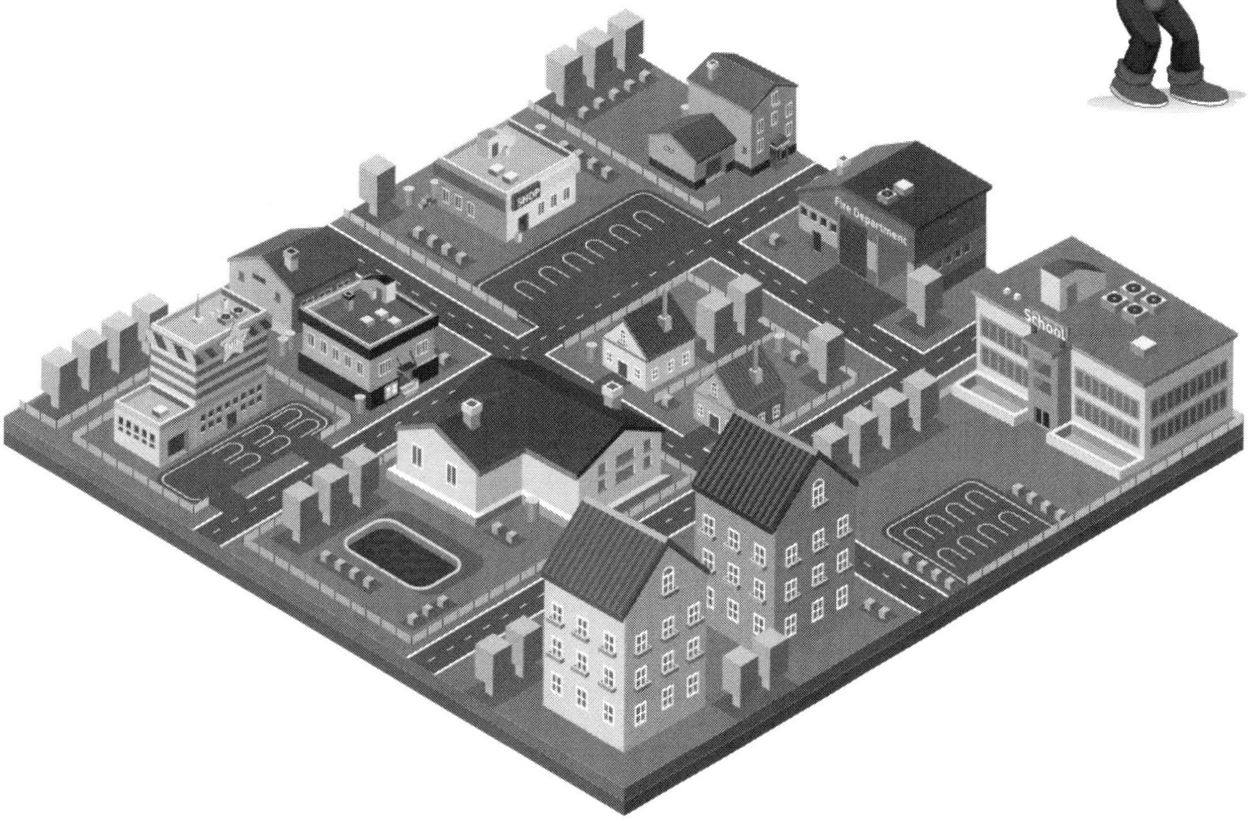

a. What is the actual area of the neighborhood? Represent using the place value chart.

b. If the area of the park in the neighborhood is 1/10 times the total area of the neighborhood, what is the actual area of the park?

Exponential and Standard Form

a. Write the following in exponential form

Standard Form	Exponential Form
10,000	
1,000	
10 x 1000	
1,000 x 1,000	
100,000	

b. Write the following in standard form

Exponential Form	Standard Form
10^3	
$8700 \div 10^4$	
3.808×10^2	
$76 \div 10^4$	
9×10^5	

Number Patterns

Complete the patterns below.

a.

	6.3	63	630		

b.		9876	987.6		9.876	

c.	7,560,000	75,600		7.56	

d.	0.009	0.9	90			

Help Detective Jones

Use the clues given below to solve.

a. One billion is equal to

b. One million is equal to

c. The missing digit in 546,889; 546,989; 547,*89

d. Which is the greatest number?

e. Which is the smallest number?

The clues are color coded and are in the same order as the riddle. Circle the correct answer to help him solve the riddle.

10 millions	1000 millions	1000 hundreds	10,000 hundreds
1/10th of a billion	1/100th of a billion	1000 times thousands	100 times hundreds
3	2	1	0
1,100,001	1,110,110	111,111	1,100,111
999,989	998,899	999,888	989,989

Solve the following multiplication problems:

a. 3000 x 500

b. 10 x 9000

c. 506 x 8

d. 24 x 650

e. 7005 x 67

f. 3756 x 14

Complete the table. One is done for you.

Words	Equation	Solution
20 times the sum of 12 and 3	**20 x (12 + 3)**	**300**
	(1400 + 1300) ÷ 100	
	14 x (567 - 230)	
Divide the difference between 1236 and 1200 by 3		
23 times the sum of 149 and 120		

Aaron's Having a Party

Aaron is having a party at his home. He buys 42 bags of chips and 36 cans of juice. Each bag of chips costs $3.45 and a can of juice costs $1.25. Use the standard algorithm to show how much he spent for the refreshments?

Chips 	Juice 
Total:	

Ryan's Math Assignment

Ryan is working on his math assignment. Look at the problems and help him answer the questions.

a. 48 x 50 = ________ b. 4.8 x 50 = ________

What is the similarity between the two answers?	
What is the difference between the two answers?	
Explain your thinking:	

The Missing Divisor

Find the missing divisor. Express the answer in exponent form.

a.

5.2

Exponent form: ← ☐) 52

b.

45

Exponent form: ← ☐) 4500

c.

4.567

Exponent form: ← ☐) 4567

d.

7.689

Exponent form: ← ☐) 76.89

Draw 😃 if the equation is true or ☹ if the equation is wrong. And write your reasoning.

Equation	Or	Reasons
a. 5 tens = 3 tens + 2 tens		
b. 53 x 100 x 8 = 530 x 8 x 10		
c. 67 x 90 x 100 = 67 x 9 x 10 x 10 x 10		
d. 50 tens x 6 hundreds = 60 tens x 5 hundreds		
e. 5 x 78 x 3000 = 50 x 780 x 3		

Solve the following.

a. A number when multiplied by 100 is 645,300.

What is the number?

b. 42.35 becomes 0.4235 when _______ (multiplied/divided) by ________

c. 30 rows of 10 inch long tiles will be of length _____ inches.

d. A number when divided by 1000 is 0.0985.

What is the number?

How Much Will Angela Save?

Angela deposits $540 every month in her bank account.

a. Estimate how much will be in her account by the end of the third year?

b. Approximately how long will it take for her to save $36,000?

Miranda is in charge of making cookies for the class party.

a. Each batch of cookies needs 0.5 cups of cocoa powder. If she is making 14 batches, how much cocoa powder will she need?

b. Miranda has mixed the following dry ingredients to make one batch of cookies. What is the total weight of the dry ingredients? Look at her recipe card to answer.

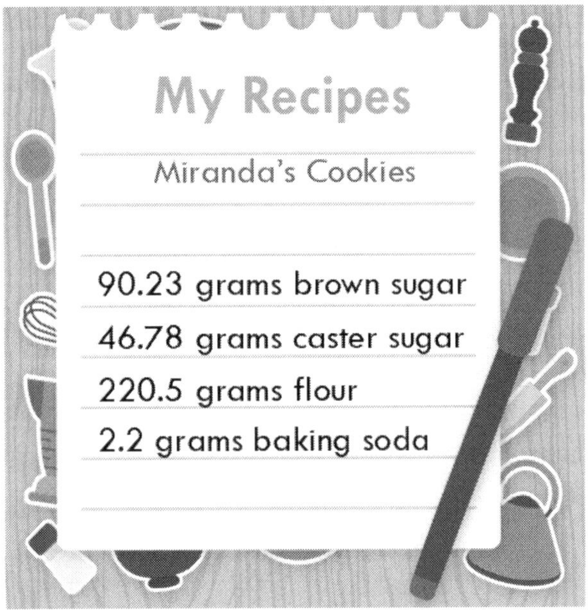

c. The first batch of cookies weighs 360.58 grams and the second batch weighs 297.89 grams. Which batch weighs more and by how much?

d. She has 480.95 grams of chocolate chips. She wants to spread them across 5 batches. How much chocolate chips will each of the five batches get?

e. Miranda uses 16.9 oz of vanilla essence to make her cookies. If she has to multiply the amount of cookies by 100 times, what will be the amount of vanilla essence needed?

Solve the multiplication problems using the grids below. Each square in the grid represents 1 hundredths.

Directions:

- Shade one factor vertically and the other horizontally.
- The overlapping area is the product of the two factors.

The first one is done for you.

a. 0.3 x 0.2
 = 0.06

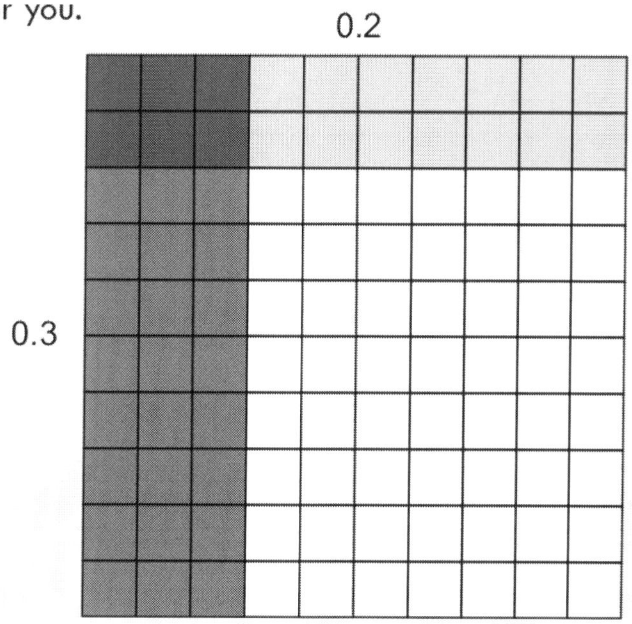

b. 0.8 x 0.1

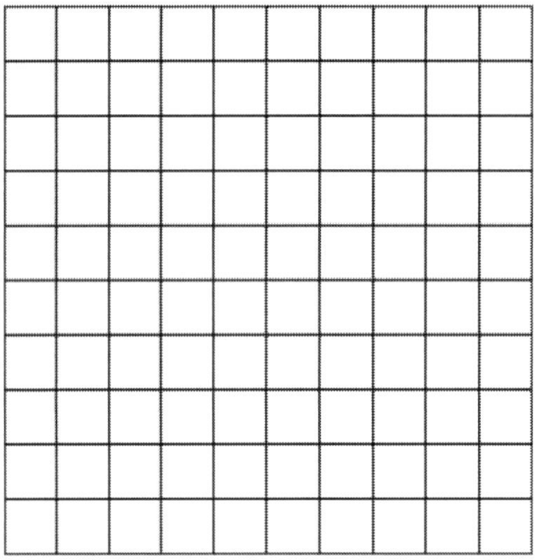

c. 0.4 x 0.6

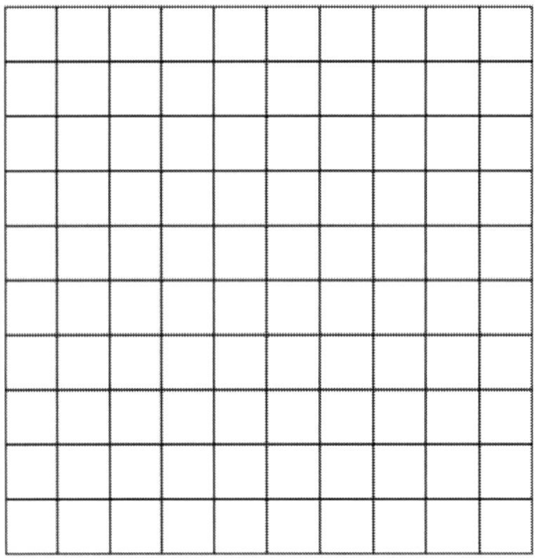

d. 1.3 x 0.2

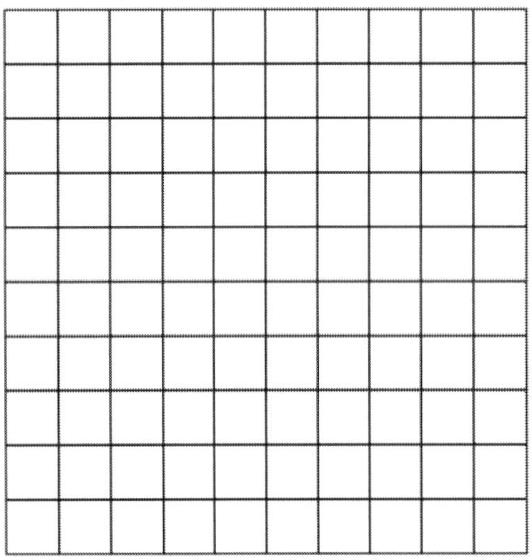

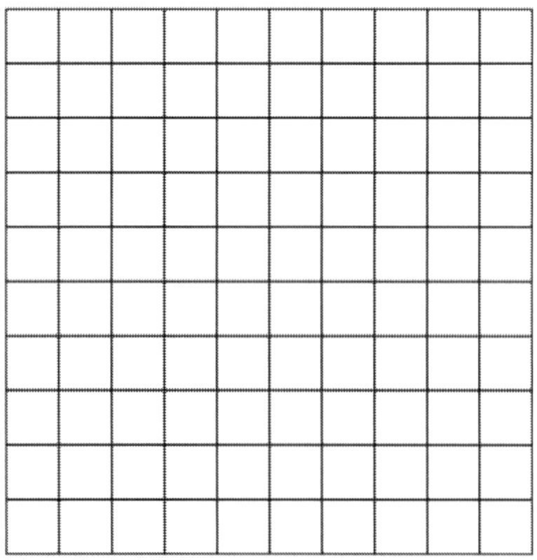

Number operations in fractions

Express each of the following as a sum of a whole number and a fraction.

a. $\frac{9}{7}$
b. $\frac{32}{7}$
c. $\frac{7}{2}$
d. $\frac{24}{9}$

Equivalent fractions

Express each fraction as the sum of two or three equal fractional parts.

a. $\frac{6}{7}$
b. $\frac{15}{4}$

Fraction on Number Line

Show each fraction on a number line and then solve.

a. $\frac{1}{4} + \frac{1}{4} + \frac{1}{4} + \frac{1}{4}$
b. $\frac{1}{9} + \frac{4}{9}$
c. $\frac{2}{7} - \frac{2}{7}$
d. $\frac{3}{5} - \frac{1}{5}$

Fraction Model

Draw a rectangular fraction model to find the sum or difference of each.

a. $\frac{1}{2} + \frac{2}{3}$
b. $2 + 1\frac{1}{5}$
c. $1\frac{3}{4} + \frac{1}{5}$

d.	$\frac{5}{6} - \frac{3}{4}$
e.	$1\frac{1}{3} - \frac{1}{2}$

Greater Than or Less Than 1

Add the fraction and shade the answer as greater than or less than 1.

a. $\frac{1}{4} + \frac{1}{2}$	Greater than 1	Less than 1
b. $1\frac{1}{5} + \frac{1}{3}$	Greater than 1	Less than 1
c. $4\frac{5}{8} + 3\frac{3}{5}$	Greater than 1	Less than 1

Rearrange the terms and then add or subtract the fractions.

a. $\frac{1}{4} + 2\frac{2}{3} + \frac{7}{4} + \frac{1}{3}$

b. $\frac{5}{6} - \frac{1}{6}$

c. $4\frac{3}{7} - \frac{3}{7}$

d. $2\frac{3}{5} + \frac{2}{5}$

Make the Statements True

Complete these equations to make the statement true.

a. $11\frac{2}{5} - 3\frac{2}{5} =$ ____

b. $\frac{5}{12} -$ ____ $= \frac{2}{12}$

c. ____ $- 7\frac{1}{4} = 23\frac{1}{4}$

Solve the given word problems.

a. Kim cut out 5 wooden boards of equal length. Each was nine tenths of a meter long. What is the total length of all boards put together? Express the final answer as a mixed fraction.

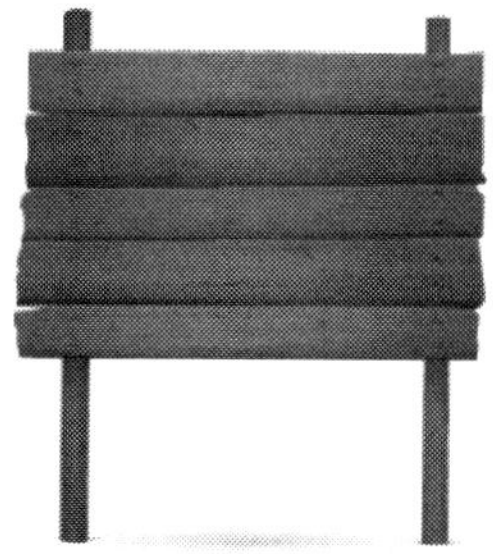

b. Nancy took $1\frac{1}{4}$ hour to complete a marathon whereas Tom took $\frac{8}{9}$ hour. How much longer did Nancy take than Tom?

c. Carl exercised for 30 minutes. He worked his core muscles for $24\frac{1}{3}$ minutes and did some stretches for the remaining time. How long did he spend stretching?

d. Samuel is sewing an apron. He needs $2\frac{5}{8}$ meters of fabric for the front and $\frac{3}{5}$ meters of fabric for the strings. How much fabric does he need in total?

Justify Eric's Solution

Eric says $3 - 1\frac{1}{3}$ will be more than two since $3 - 1$ is 2. Draw a picture to show that Eric is wrong.

Is Elise Right?

The picture below shows Elise's work. Is her answer correct? Draw a picture to justify your answer.

$$5\frac{3}{4} + 2\frac{1}{4} = 7\frac{1}{2}$$

$$7\frac{4}{8} = 7\frac{1}{2}$$

Measurement

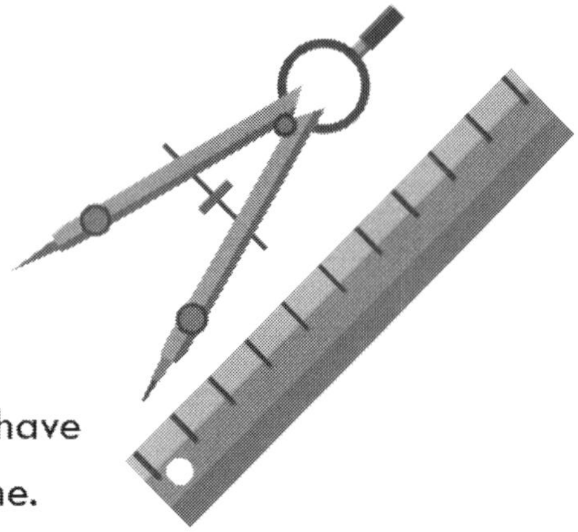

Volume is an attribute of solid-figures.

A cube with side length 1 unit, called a "unit cube," is said to have "one cubic unit" of volume, and can be used to measure volume.

A solid figure which can be packed without gaps or overlaps using 'n' unit cubes is said to have a volume of 'n' cubic units.

Let's practice

Mr. Rod Has only a Centimeter Tape.

Mr.Rod is a carpenter and is constructing a dining table. The dimensions are in meters, centimeters, and millimeters, but he has only a centimeter tape. What do you think he can do? Help him by converting them into a suitable unit.

Convert to Like Units

Use simple operations to solve these word problems. Convert them into like units wherever needed.

a. Jean poured 1l of lime juice in a jug. She then added some 250 ml of soda. What is the quantity of lime soda in the jug?

b. Four people in a relay took 165 seconds to reach the finish line. How many minutes did they take?

c. Kristy cut a 4.8 m long rope into 10 pieces. Olivia cut a 570 cm long rope into 10 pieces. How much longer is Olivia's rope piece than Kristy's?

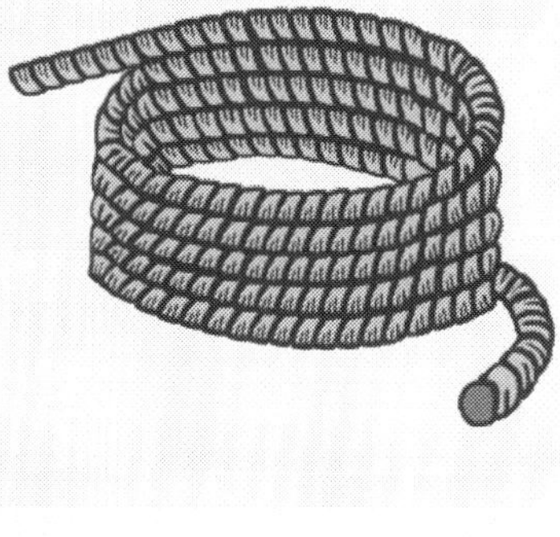

Compare the Measurements

Compare these measurements using the symbols > , <, or =.

a.

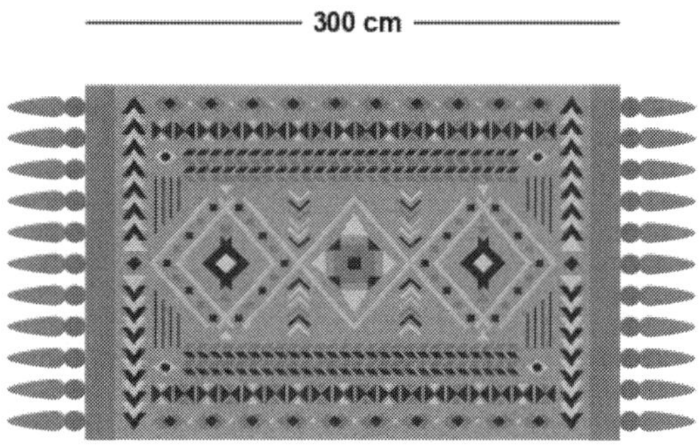

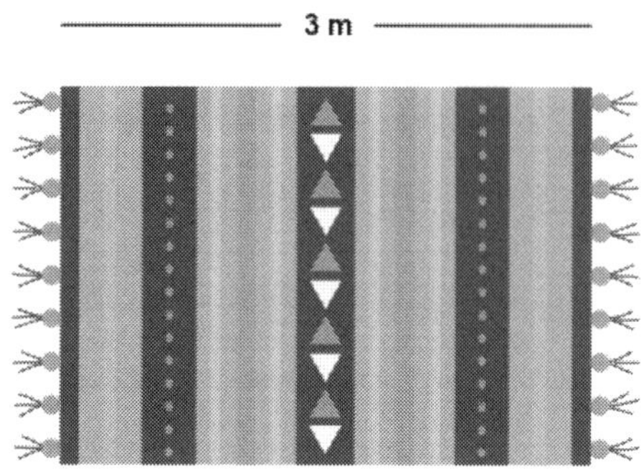

b.

c.

Water Consumption of Miley and Lilly

The table shows the one day consumption of water by Miley and Lilly. Answer the questions by looking at the table.

Activity	Amount of water used by Miley	Amount of water used by Lilly
Car Wash	25 gallons	120 quarts
Brushing teeth	2 gallons	4 quarts
Taking a shower	40 quarts	15 gallons
A load in the washing machine	30 gallons	25 gallons
Flush a toilet	7 gallons	10 quarts

a. What is the total amount of water used by each of them?

b. Who uses more water to wash their car?

c. Later, Miley learnt that by closing the tap while brushing, she could save $\frac{2}{3}$ units of water. Is she still using more water than Lilly?

How Much Ribbon?

Lisa decided to make some scarves for an exhibition. She needs 1.5 yards of ribbon for each scarf.

a. At the fabric store, ribbon is sold by feet. If she wants to make 25 scarves how many feet of ribbon will she need?

b. If the ribbon costs 10¢ per foot. How much would she have to pay in dollars?

c. Marty's Apparel decided to make 100 times as many scarves as Lisa. Write an expression to show the amount of ribbon the company needs to make the scarves.

Volume of Solid Shapes

Demi says that the given figure, made of 1 cm cubes, has a volume of 5 cubic cm. Is she right? Explain your answer.

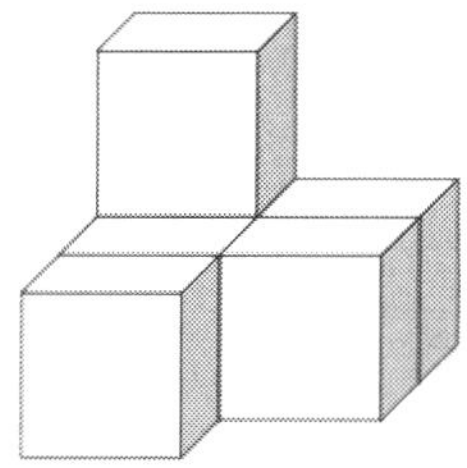

Find the Volume

Find the volume for each of these figures.

a. Stack of wooden blocks of the same size.

b. Stack of blocks that are 1 cubic cm each.

c. A Rubik's cube with cubelets that measure 1 cubic cm each. There are 27 cubelets in a Rubik's cube.

Draw a structure on the dot paper for the given volumes.

a. 7 cubic units.

b. 4 cubic units

Predict the Number of Cubes

Predict how many centimeter cubes will fit in the black box for each image.

a.

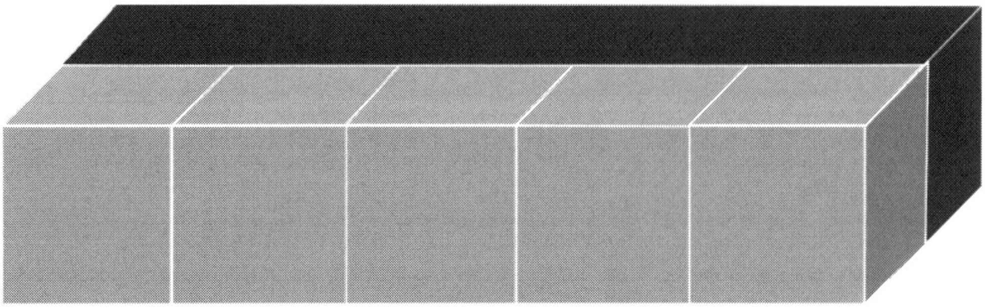

b.

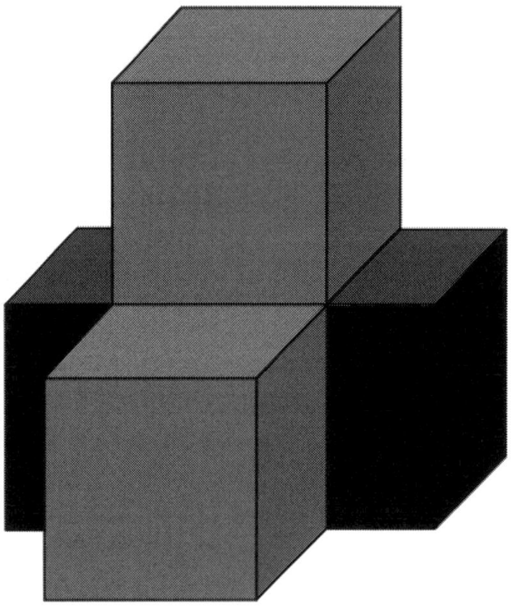

Guess the Number of Cubes

Emma hands out a net template to her friend and asks her to fold it to make a box. She then asks her to guess the number of unit cubes that would fit into the box. Help Emma's friend find the answer.

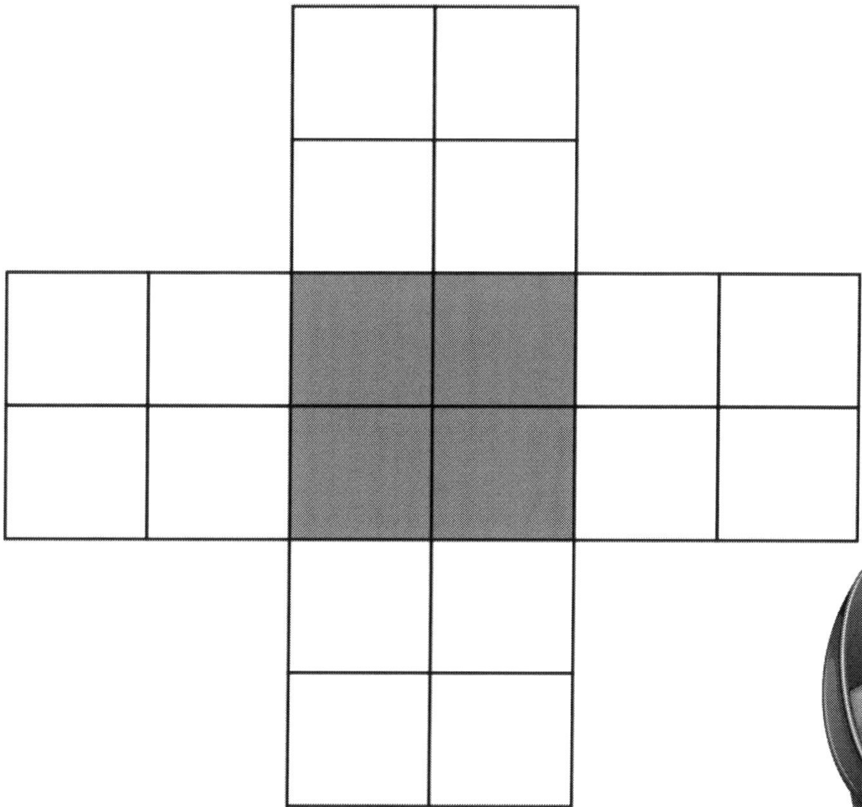

Robert and Jamie's Argument

Robert and Jamie have calculated the volume of this prism. Both the boys agree that there are 4 layers. Read their conversation and explain who is right.

Olivia makes a prism using 1-inch by 5-inch by 5-inch cubes. She then added layers to this prism. Complete the table with the volume and provide an explanation of how you arrived at the volume of each prism.

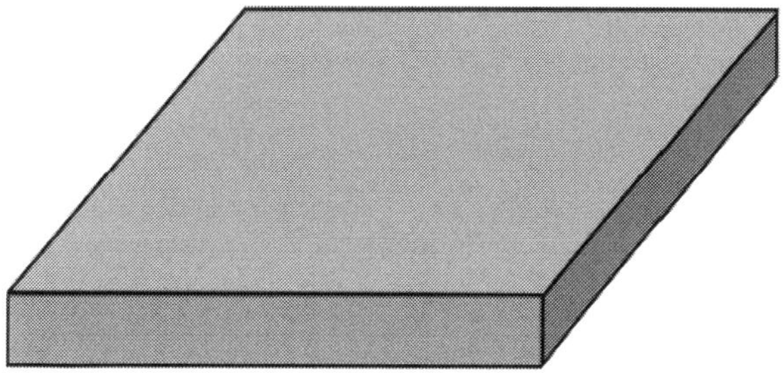

Total number of layers	Volume	Explanation
2		
5		

Calculate the volume of each prism. Write the units for each answer.

a.

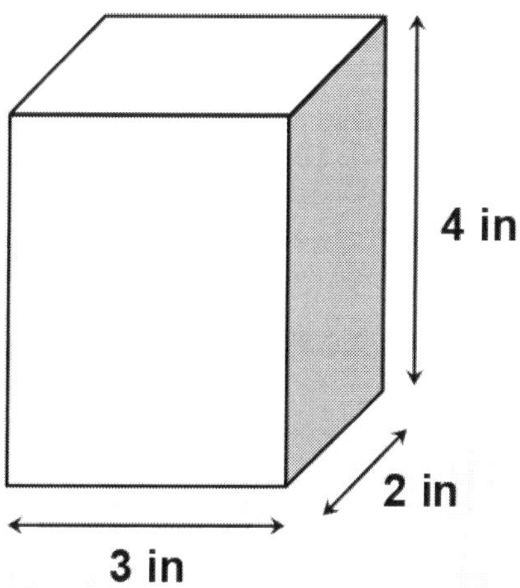

b.

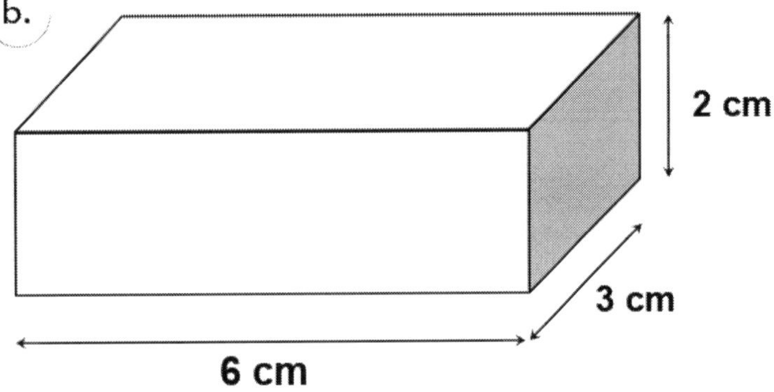

c.

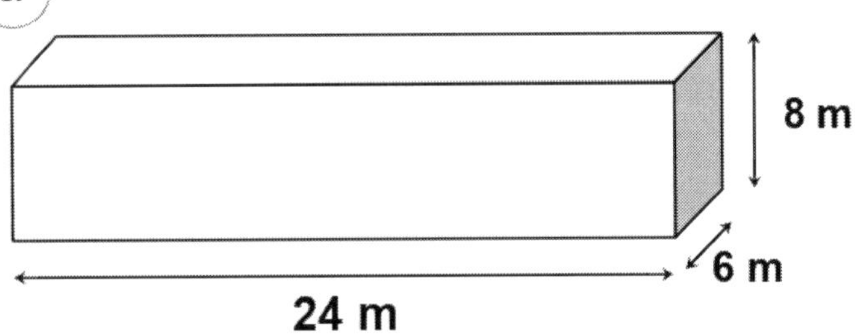

d.

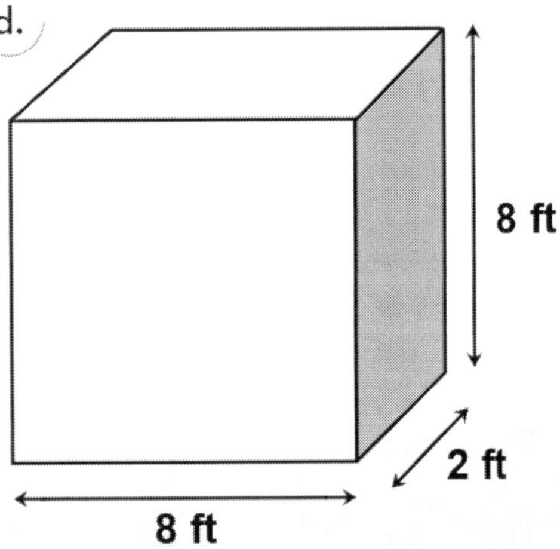

Calculate the volume in the given problems.

a. Mr. Paul is constructing a cupboard in the shape of a rectangular prism. It has a length of 28 inches, a width of 24 inches, and a height of 36 inches. Find the volume of the cupboard.

b. A storage box has a surface area of 56 square centimeters and a height of 4 centimeters. Calculate the volume of the box.

c. A container is 25 cm long and 20 cm wide. If it holds 1 liter of water when it is full, what will be the height of the container?

d. Dean decides to build a box for his art supplies. He has a paint box that is 12 inches wide that he wants to lay flat at the bottom of his box. He wants his box not taller than 2 inches. Help him find one way to build a box that has a volume of 72 cubic inches.

Find the total volume of these shapes put together.

a.

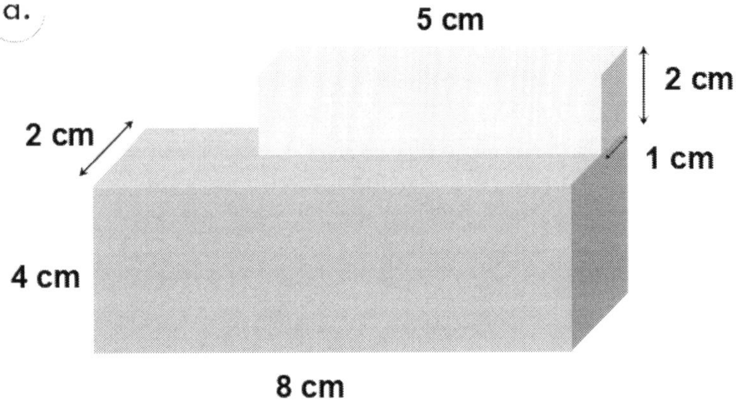

b.

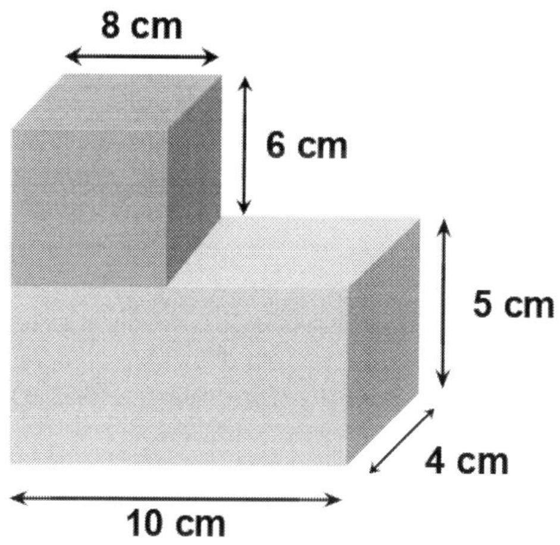

c.

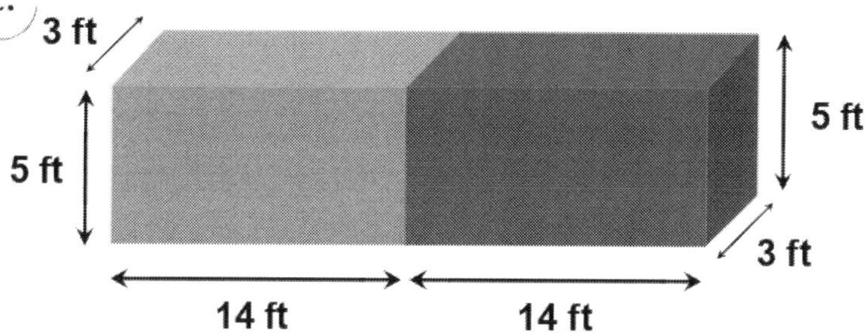

Calculate the volume of the prism.

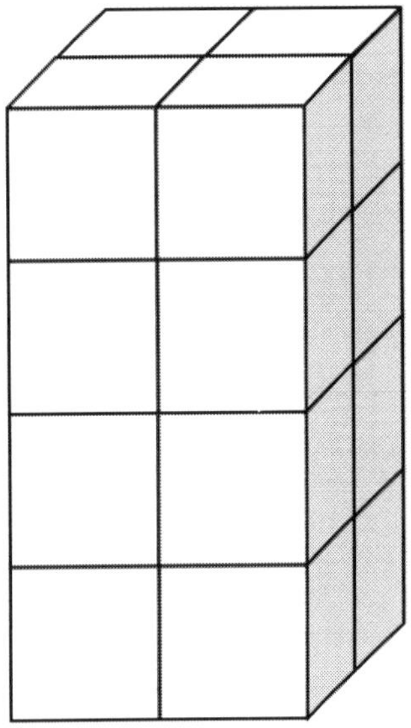

Length: ________ mm

Width: ________ mm

Height: ________ mm

Volume: ________ mm^3

Write the multiplication statement that shows the calculation of the volume.

Line Plots

Have you used a Thesaurus? It is similar to a dictionary, but it helps you find synonyms for a particular word instead of meanings. Using synonyms will keep us from repeating the same word often.

Let's practice

Fund-raising Rally Plot

Long Valley School organized a fund-raising rally.

Use the line plot below to answer questions.

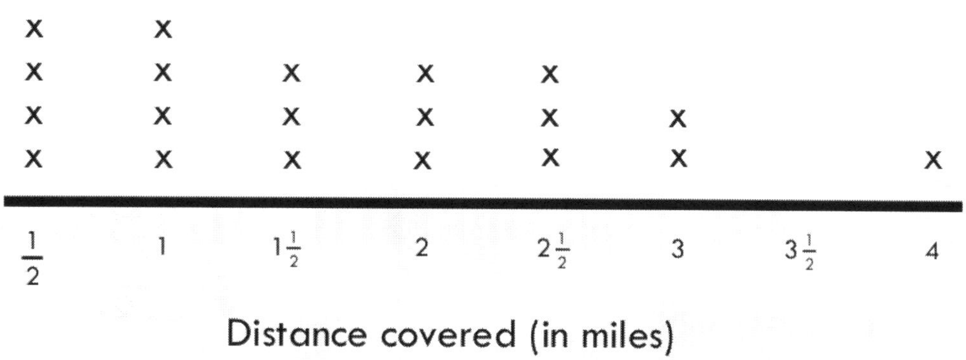

a. How many students rode for more than 2 ½ miles?

b. How many students rode for less than 1 mile?

c. How many students participated in the rally?

d. What is the total distance planned for the rally?

e. How many students completed the rally?

The line plot shows the lengths of all the pieces of ribbon Kate used for an art project. She cut all the pieces from one original piece of ribbon.

Length of the ribbon pieces (in inches)

a. How many pieces of ribbons does she cut?

---

b. What is the minimum and maximum length of the pieces?

---

c. What is the length difference between the longest and shortest pieces?

---

d. If she cuts one 2 ½ inch piece into 1-inch pieces, how many 1 inch pieces will she have?

---

e. What is the total length of the ribbon before Kate cut them into pieces?

Laura's Reading Schedule

Create a line plot for the given set of data. Then answer the questions that follow.

Hours Laura spent reading each day last month.

1, 1 1/4, 1
1, 1 1/4, 1 1/4,
1 1/2, 1 3/4,
2, 1,
1 1/4, 1 3/4,
1 1/2, 1,
1 1/4,
1, 1 1/4,
1 1/2, 2,
1, 1 1/4, 1 3/4,
1 3/4, 1 1/2, 1

a. What was the most amount of time Laura spent reading in one day?

____

b. What was the least amount of time Laura spent reading in one day?

____

c. How many days did Laura spend reading for exactly 1-hour?

____

d. How many days of reading has been recorded for data?

____

e. How many hours did Laura spend reading last month altogether?

____

Draw line plot here:

Did Sean Collect Enough Water?

Sean collects the water from the AC vent and reuses the water for his garden fountain.

He needs at least 3 gallons of water for his fountain. He collects water every one hour and the data is represented on the line plot.

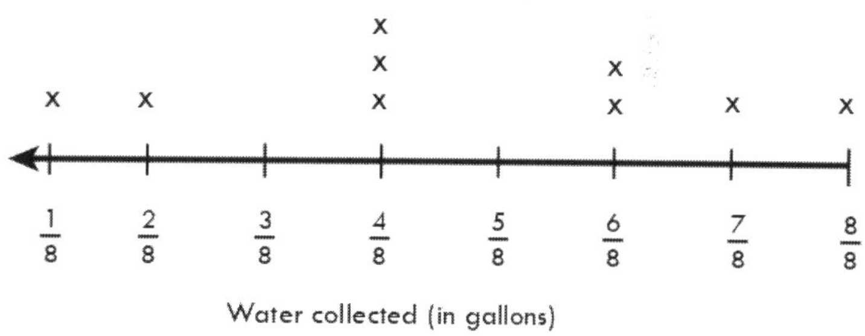

Look at the line plot to answer the question. Did he collect enough water to reuse?

Fair Share of Pudding?

Leo and his 7 friends brought some pudding. Each of them have different amounts of pudding right now. They want to be fair, so everyone should get the same quantity of pudding.

They represented the amounts on a line plot.

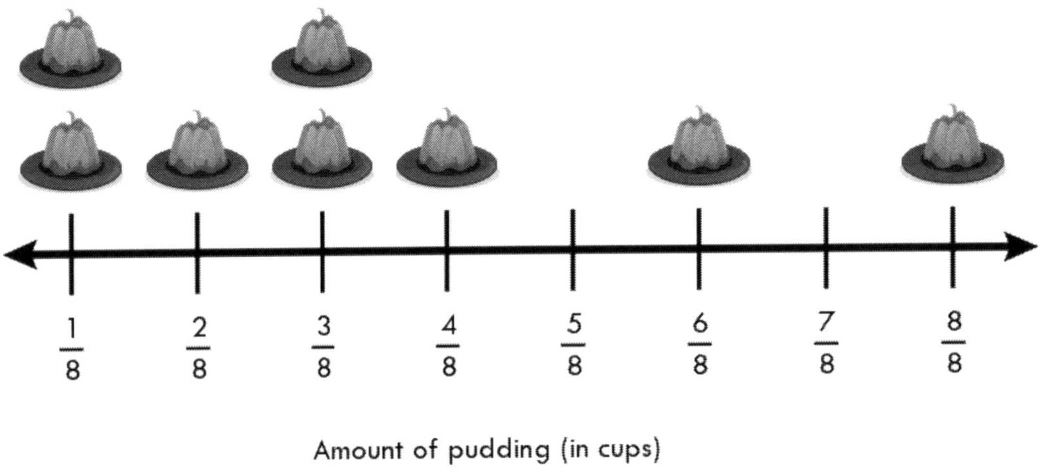

a. How can they redistribute so that everyone gets the same amount? How much pudding will each get after redistribution?

Geometry

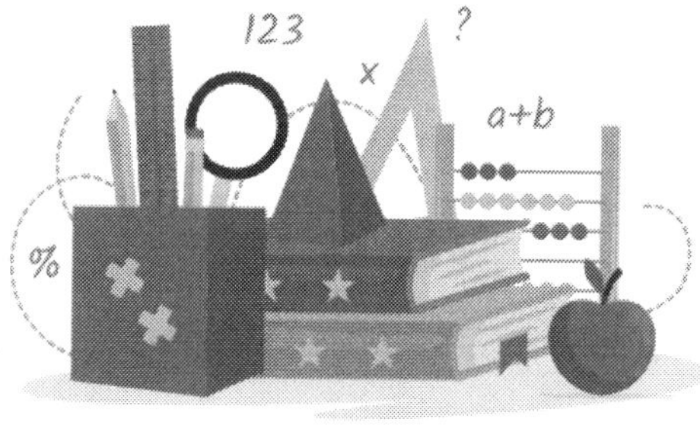

A coordinate system is defined by a pair of perpendicular number lines called axes, with the intersection of the lines (the origin) arranged to coincide with the 0 on each line and a given point in the plane located by using an ordered pair of numbers, called its coordinates.

Let's practice

Draw a pair of parallel lines in each box.

Then, use these lines to draw a trapezoid as mentioned.

Finally, shade a pair of angles that equals to 180° in each of the trapezoids drawn.

a. No right angle

b. 2 obtuse angles

Follow the instructions to answer (a) and (b)

Step 1: Draw a line XY parallel to this page.

Step 2: Draw two 50° angles with X and Y as the vertices.

Step 3: Extend the line until they intersect, and label this as Z.

Do your drawing here:

a. Name the shape that you drew.

b. Draw a line inside the shape such that you get one trapezoid. Shade the trapezoid.

A rhombus has a perimeter of 217 cm. What is the length of each side of the rhombus?

Draw a kite that is not a parallelogram on the grid paper.

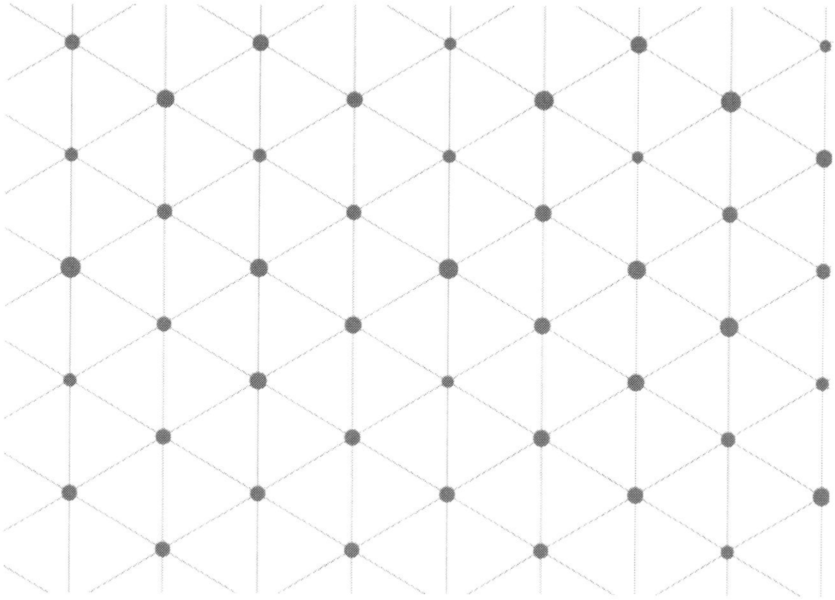

a. List all the properties of a kite.

b. When can a parallelogram also be called a kite?

Zoe says that the figure ABCD is a quadrilateral, since it has 4 points and 4 segments. Is he right? Explain your answer.

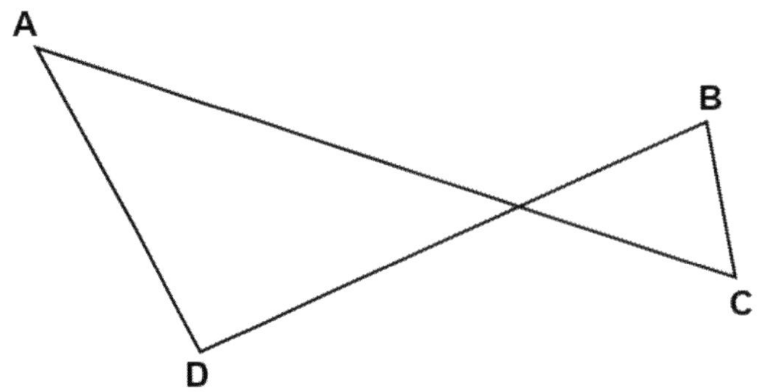

Eric and Summer have a competition among themselves on naming the properties of a rhombus. Eric got 2 correct and Summer got 3. Can you think of more? List all the properties that you know.

Can You Convert a Rectangle Into a Square?

How can you convert the given rectangle into a square? Write down what you changed and then draw the square in the space provided.

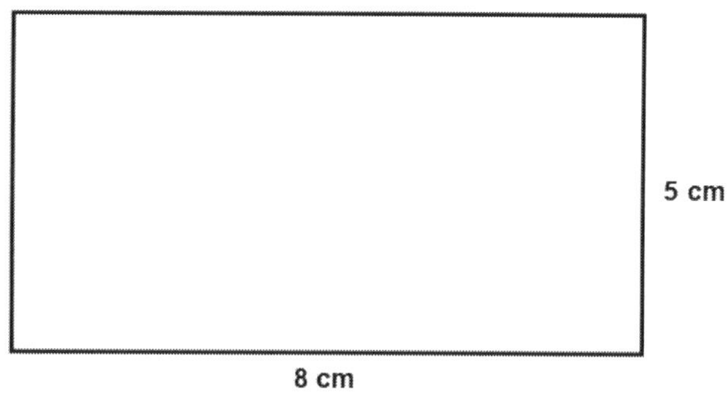

I changed -

Draw your square:

Geometry Quiz

Take this quick quiz on two-dimensional figures.

a. What is an isosceles trapezoid?

____

b. Name one unique attribute that distinguishes a trapezoid from a parallelogram.

____

c. When is a rectangle also called a rhombus?

____

d. What is a rhombus with 4 right angles called?

____

Use the words in the box to fill in the blanks.

rectangles	parallelograms

a. All ____________ are ____________ but all ____________ are not ____________ .

kites	rhombuses

b. All ____________ are ____________ but all ____________ are not ____________ .

∢P is 60°. Extend the rays of P to draw a parallelogram PQRS. Measure ∢Q ∢R and ∢S. Using this, write two attributes of this parallelogram.

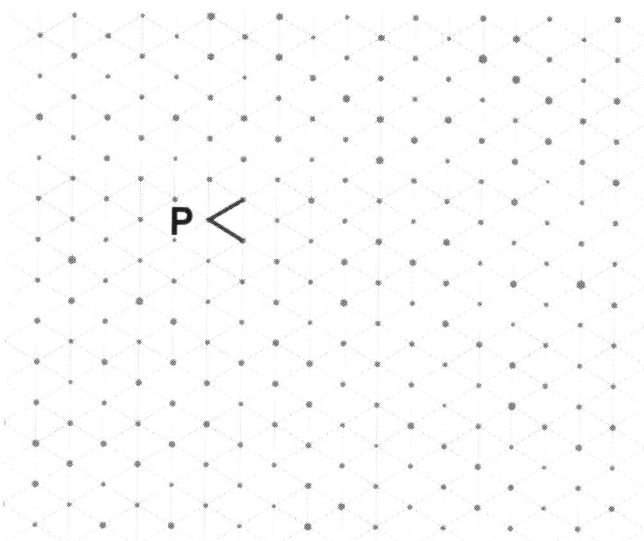

Attributes:

Function Table

Complete the function table.

X: Add 2 to get the next term

Y: Add 4 to get the next term

X	0			
Y	0			

X: Add 3 to get the next term

Y: Add 6 to get the next term

X	0
Y	0

X: Add 5 to get the next term

Y: Add 1 to get the next term

X	0
Y	0

X: Add 3 to get the next term

Y: Add 2 to get the next term

X	0
Y	0

Write the ordered pairs for each image.

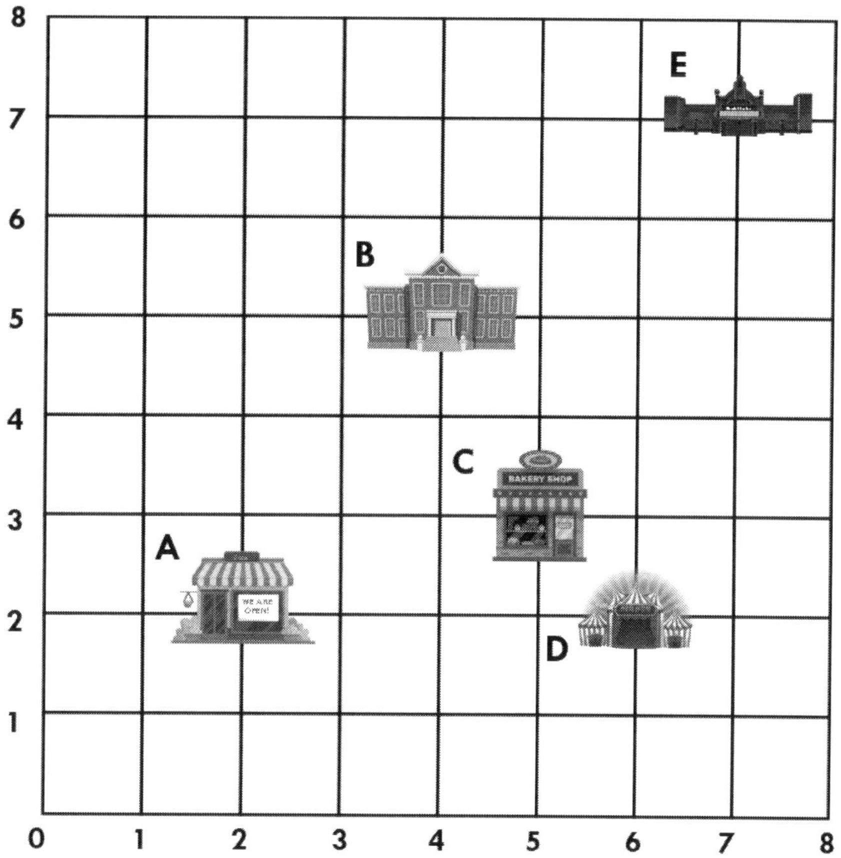

A Fruity Grid

Write the location of each fruit on the grid. One is done for you.

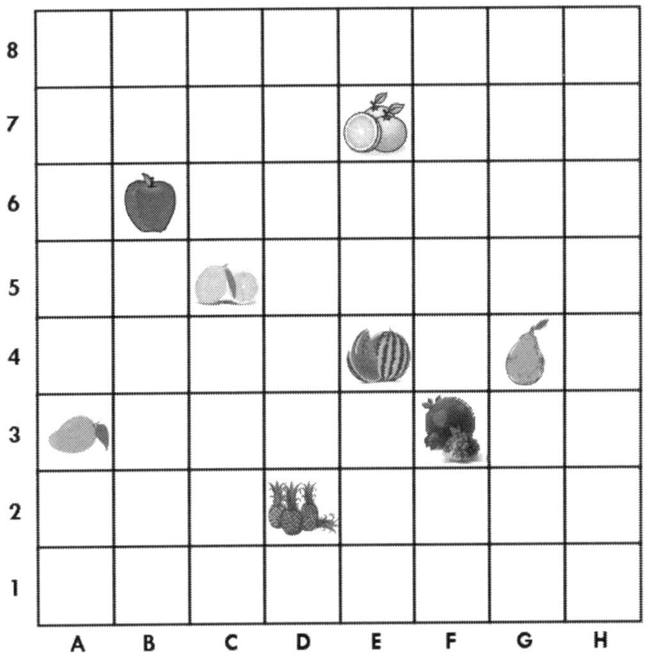

Pomegranate	**F, 3**
Apple	
Mango	
Lemon	
Pineapple	
Watermelon	
Guava	
Orange	

Use the coordinate plane below to complete the following tasks.

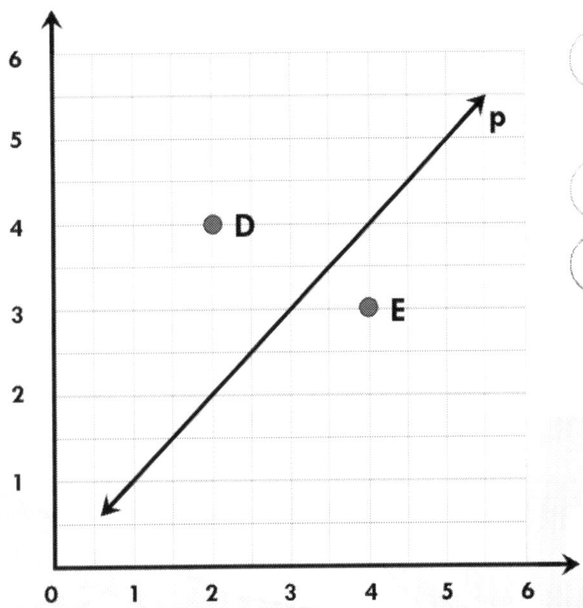

a. Construct a line, n, that is parallel to line p and contains point E.

b. Name 2 coordinate pairs on line n.

c. Identify a rule to describe line n.

d. Construct a line, m, that is parallel to line p and contains the point D.

e. Name 2 points on line m.

f. Identify a rule to describe line m.

g. Compare and contrast lines n and m in terms of their relationship with line p.

Complete the tables for the given rules.

Line m
Rule: y is three times x

x	y	(x,y)
0		
1		
2		
3		

Line n
Rule: y is three times x, and then add 1

x	y	(x,y)
0		
1		
2		
3		

a. Draw the lines m and n on the coordinate plane.

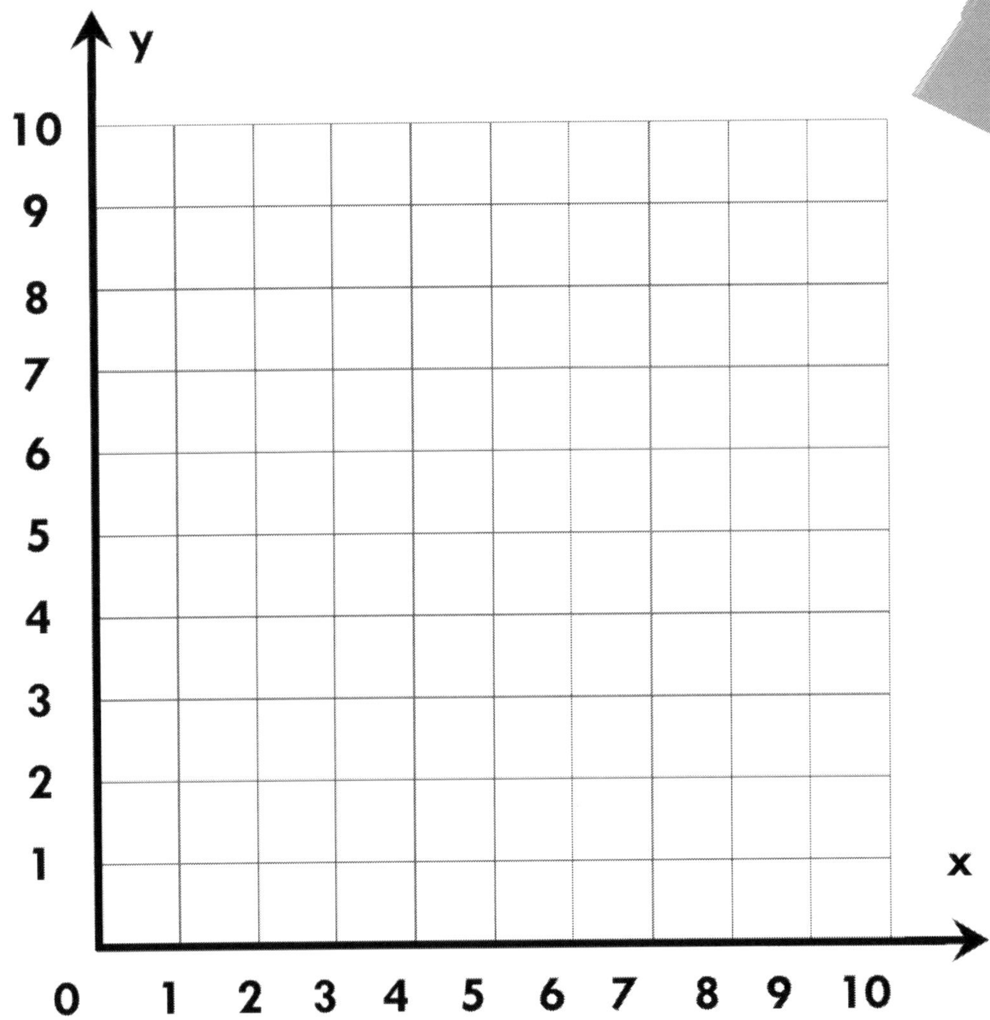

b. Compare and contrast these lines.

c. Based on the patterns you see, predict the line for the rule: Three times x minus 1. Draw the line on the plane above in red color.

Complete the table and the rule beside each table.

a.

x	0	1	2	3	___	___
y	8	9	10	___	12	___

$y = x +$ ___

b.

x	1	3	8	5	___	12
y	5	7	12	___	15	___

$y = x +$ ___

c.

x	2	3	9	5	___	___
y	8	12	36	___	24	32

$y =$ ______

d.

x	1	5	7	3	___	___
y	0	4	6	___	12	5

$y =$ ______

e.

x	4	6	8	12	___	___
y	2	3	4	___	12	8

$y =$ ______

Tour Around the City

Use the grid given below to choose the correct option.

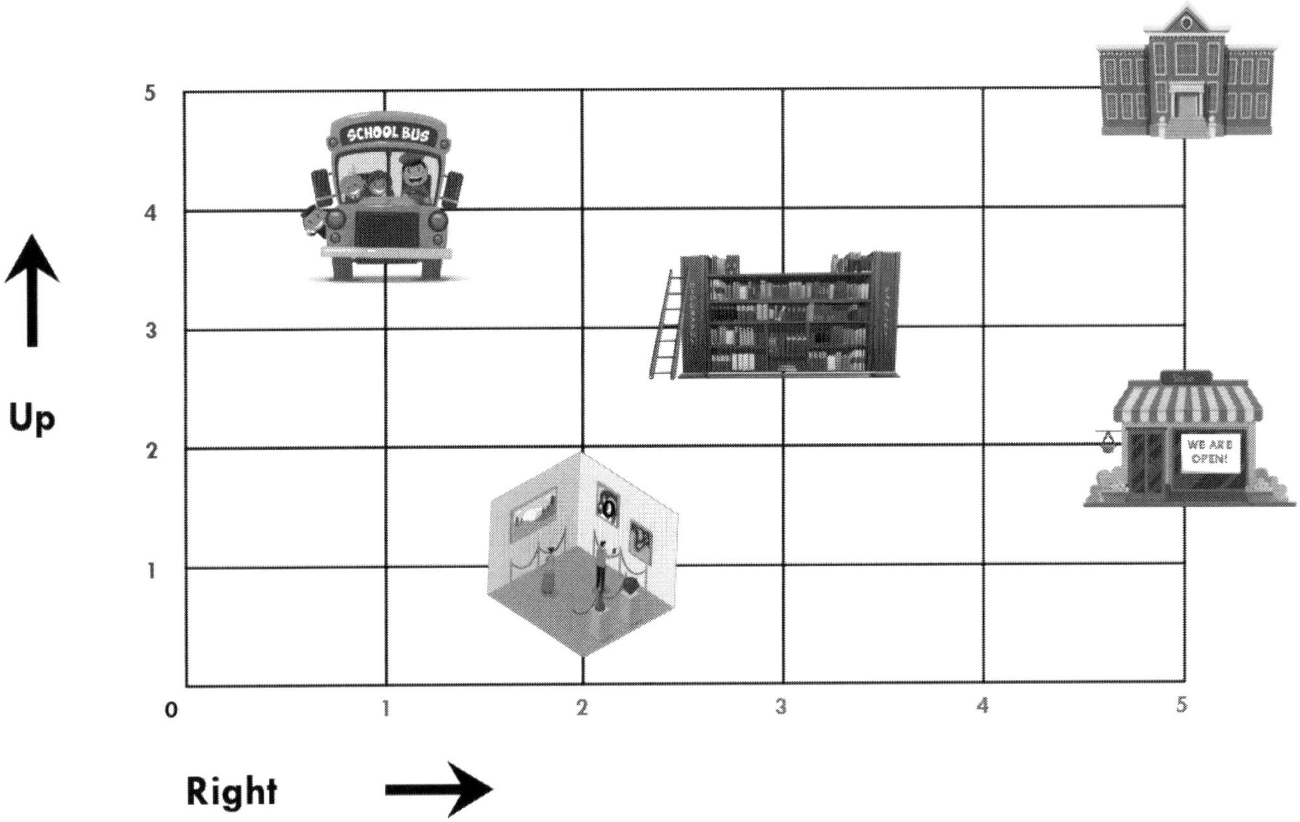

a.

- Start at 0.
- Go right 5.
- Go up 2.
- Which place did you find?

Art Gallery

School bus

Library

School

Store

b.

Start at 0.

Go right 3.

Go up 3.

Which place did you find?

Art Gallery School bus Library School Store

c.

Start at 0.

Go right 2.

Go up 1.

Which place did you find?

Art Gallery School bus Library School Store

d.

Start at 0.

Go right 5.

Go up 5.

Which place did you find?

Art Gallery School bus Library School Store

Answer each of the following questions using the coordinate plane given below.

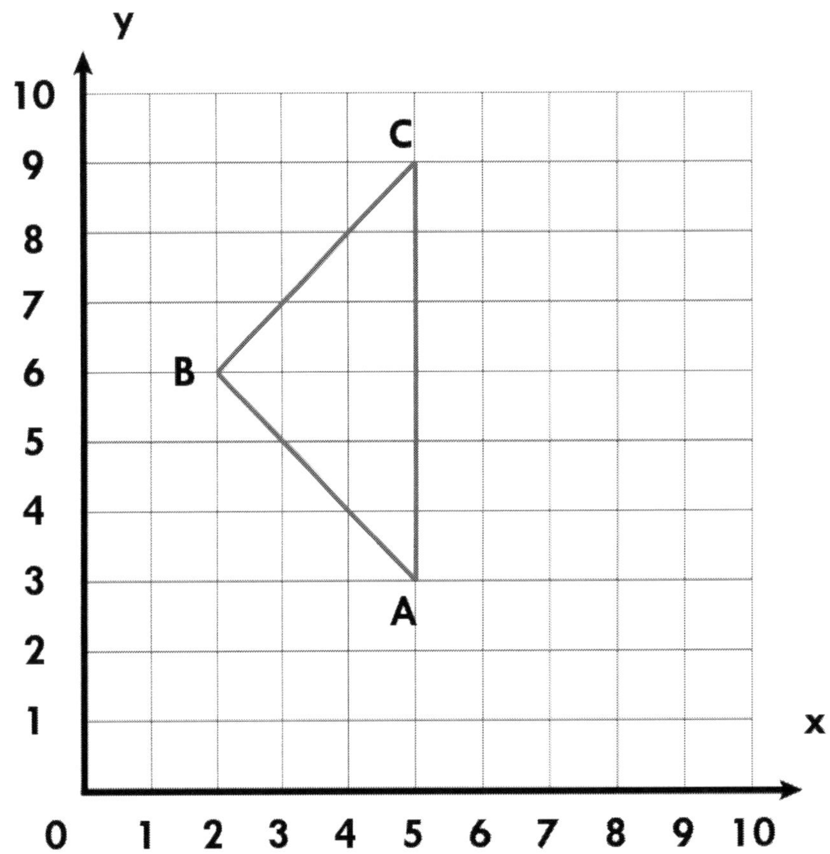

a. Which point is farthest from the origin?

b. Write the coordinates of all the vertices of the triangle ABC.

c. From the origin, go right by 8 units, and from there go up by 6 units. Mark as point D.

d. Now join the sides AD and CD. What shape is formed?

Draw a regular pentagon on the coordinate plane.

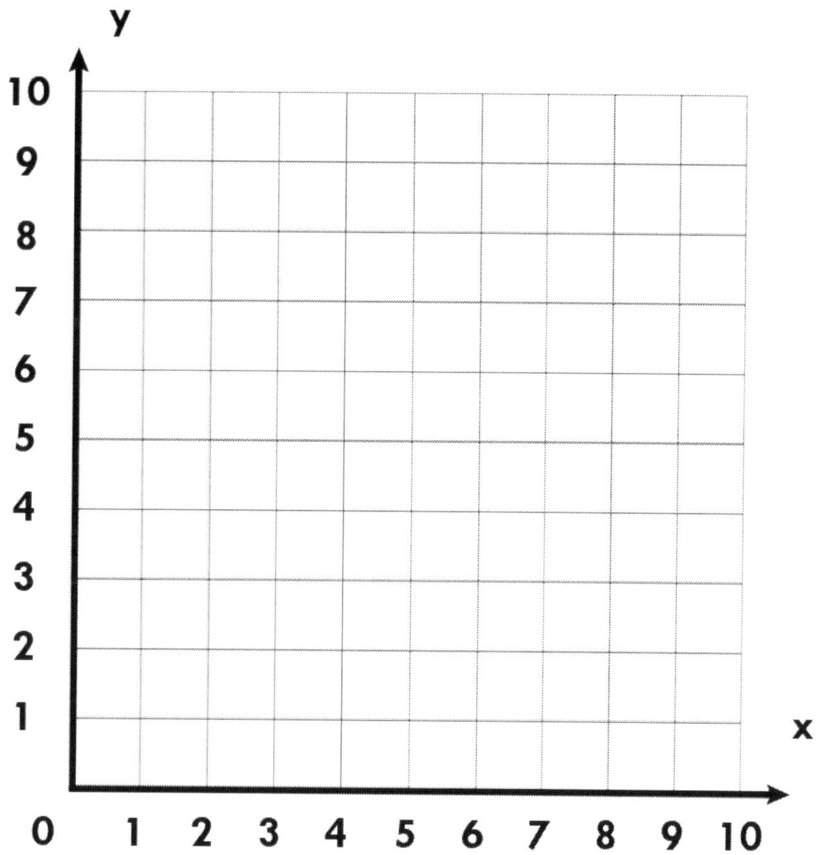

a. Mark the vertices/coordinates of the regular pentagon with letters.

b. Write the coordinates of all the vertices of the pentagon.

c. Draw a line of symmetry of this pentagon.

d. What do you think is the rule for the line of symmetry on x?

Richard's Daily Travel Plan

Richard drives 2 miles on Monday, 3.5 miles on Tuesday, 5 miles on Wednesday, 6.5 miles on Thursday, and 8 miles on Friday. Identify the rule and complete the table.

Days	Monday	Tuesday	Wednesday	Thursday	Friday	Saturday
Miles covered	2	3.5	5	6.5	8	

Rule:

Science

Help your children learn and enjoy a wide range of information and fun facts that will surprise and amaze them. Find numerous Science experiments, cool facts, activities, and quizzes for the children to enjoy learning.

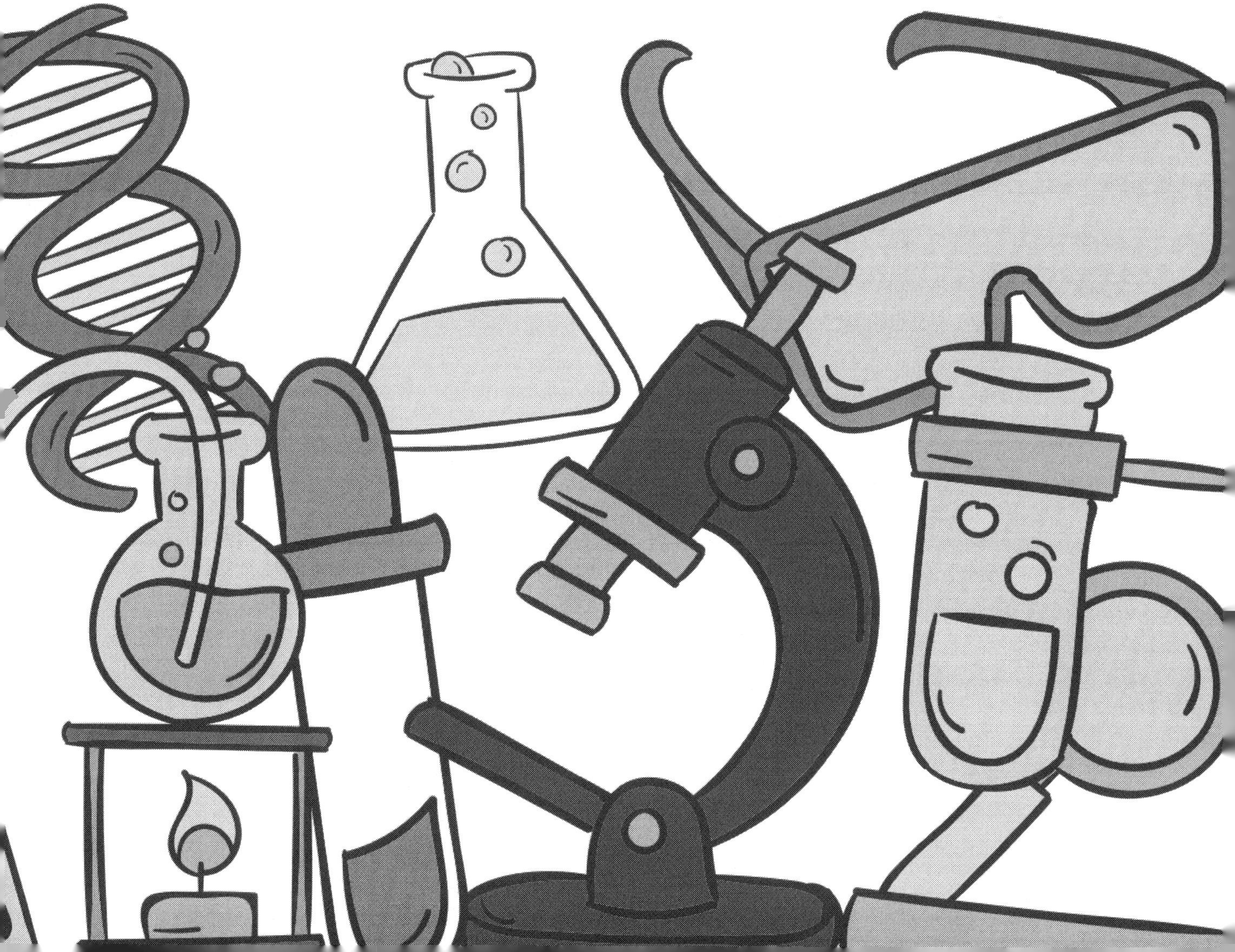

Physical Sciences

Types of Matter

Matter	anything that has mass and volume
Volume	the space taken up by an object
Mass	the amount of matter in an object
Atom	the smallest particle of an element
Element	simplest substance made of single type of atoms
Compound	substance made of two or more types of atoms
Molecule	a particle with two or more atoms joined together

objects and Elements

The objects listed below are all made up of only one type of element each. Write the name of the object and the element each object is made up of.

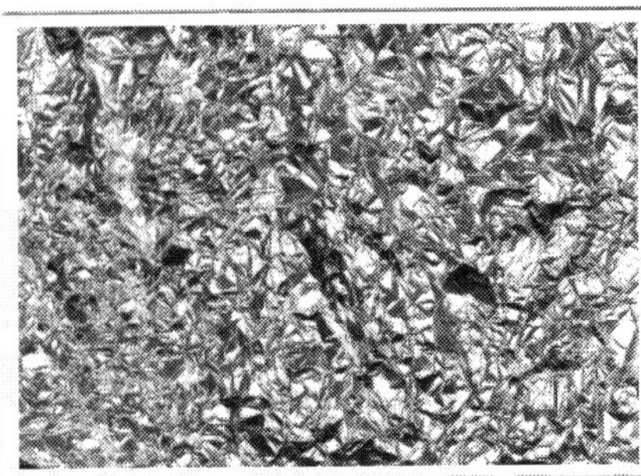		

Types of Atoms in a Compound

Identify and write the types of atoms that are present in each of these compounds.

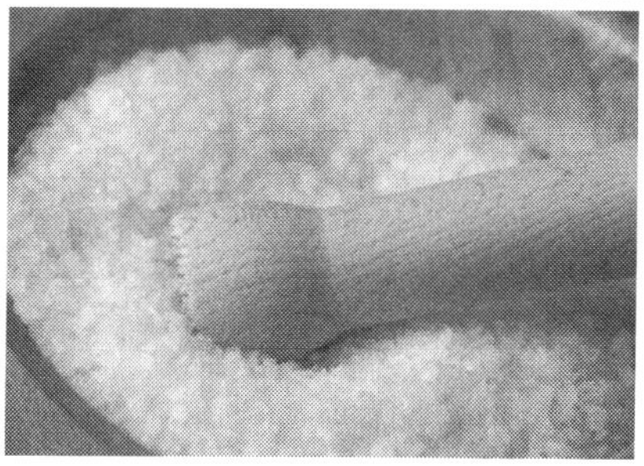 Salt	 Water	 Sugar
 Air we blow	 Mouthwash	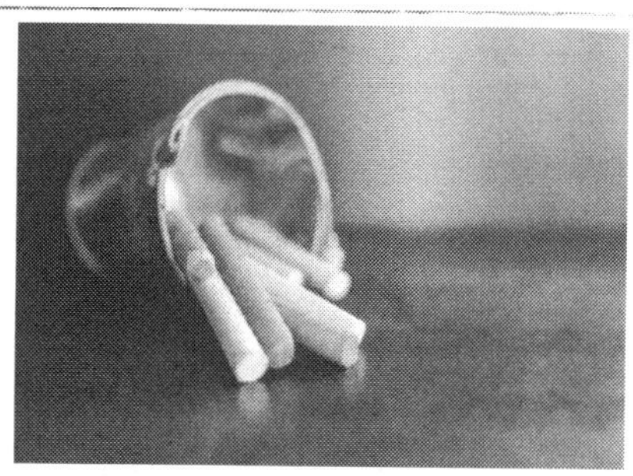 Chalk

Metals Vs Nonmetals

Define each physical property given below. Differentiate between metals and nonmetals for each of these physical properties.

	Physical Properties	Definition	Metals	Nonmetals
1.	Malleability			
2.	Sonority			
3.	Ductility			
4.	Lustre			

5.	Hardness			
6.	Thermal and electrical conductivity			
7.	Density			
8.	Melting and boiling points			
9.	Examples			

Methods of Separation

Identify the methods of separation for the given mixtures.

	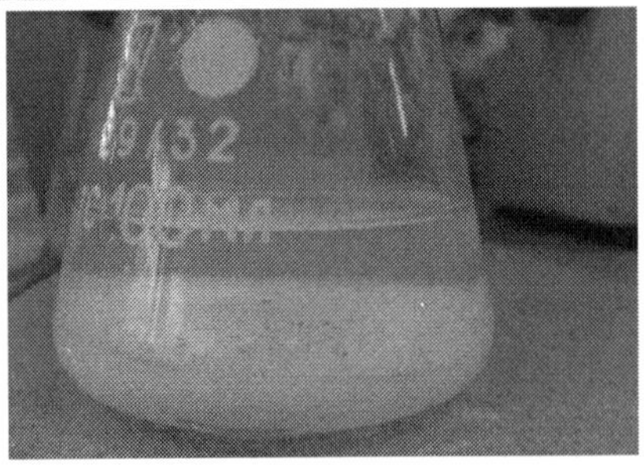	
Iron and aluminium pins	Salt in water	Two types of pulses
	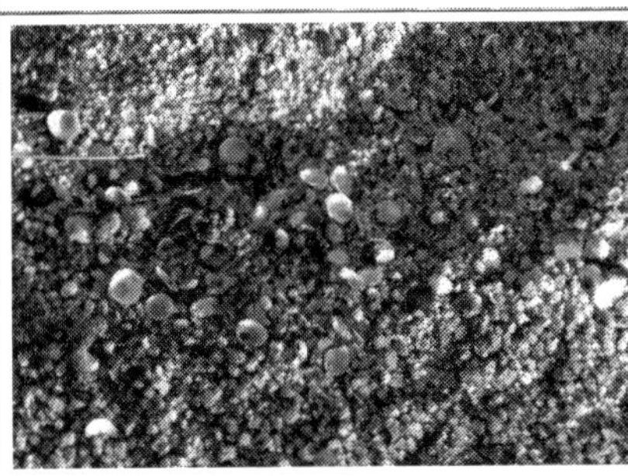	
Flour and chocolate chips	Shells from sand	Sand in water

Tea leaves from tea	Stones from a beach	Sugar in water

Crossword

Complete the crossword by answering the questions about the methods of separation.

Across:

6. The term which describes the solid trapped in a filter.

7. The term which describes the liquid that passes through a filter.

8. This is the change from solid to gas.

9. The method used to separate a mixture of sand and water.

10. The solute in a solution can be separated by the ______________ of the solvent.

Down:

1. A solute is soluble. Hence it ______________ in a liquid.

2. Apparatus used to separate a mixture of coarser and finer particles.

3. The word that describes the spontaneous mixing of gases.

4. Method of separating liquids with different boiling points.

5. Describes the liquid which does the dissolving.

Properties of Matter

Matter has two sets of properties.

Physical properties of matter can be observed with our senses can be determined without damaging or destroying the object.

Chemical properties of matter describe how they react with another substance. When a chemical property is observed, the substance is fundamentally altered.

Physical and Chemical Properties

Identify if the following are physical properties or chemical properties of matter.

No	Property	Physical Property	Chemical Property
1	Smell	☐	☐
2	Lustre	☐	☐
3	Boiling point	☐	☐
4	Flammability	☐	☐
5	Color	☐	☐
6	Density	☐	☐
7	Solubility	☐	☐
8	Supports combustion	☐	☐
9	Reacts with water to form gas	☐	☐
10	Reacts with acid to form hydrogen	☐	☐
11	Reacts with base to form water	☐	☐
12	Hardness	☐	☐
13	Taste	☐	☐
14	Melting point	☐	☐
15	Forms rust	☐	☐

Physical Vs Chemical Changes

Identify which of these are physical changes and chemical changes.

1	A balloon popping	
2	Folding clothes after they come out of the dryer	
3	When the food you eat digests in your body	
4	A towel absorbing water	
5	Folding a piece of paper	
6	Freezing water	
7	Cooking an egg	
8	Making shapes with play dough	
9	Mentos and coke combining to form an explosive reaction	
10	Dissolving sugar in water	
11	Baking cookies	
12	Clipping fingernails	
13	When gas burns in an engine to make the vehicle run	
14	When milk curdles	
15	When leaves change colour during autumn	

Periodic Table of Elements

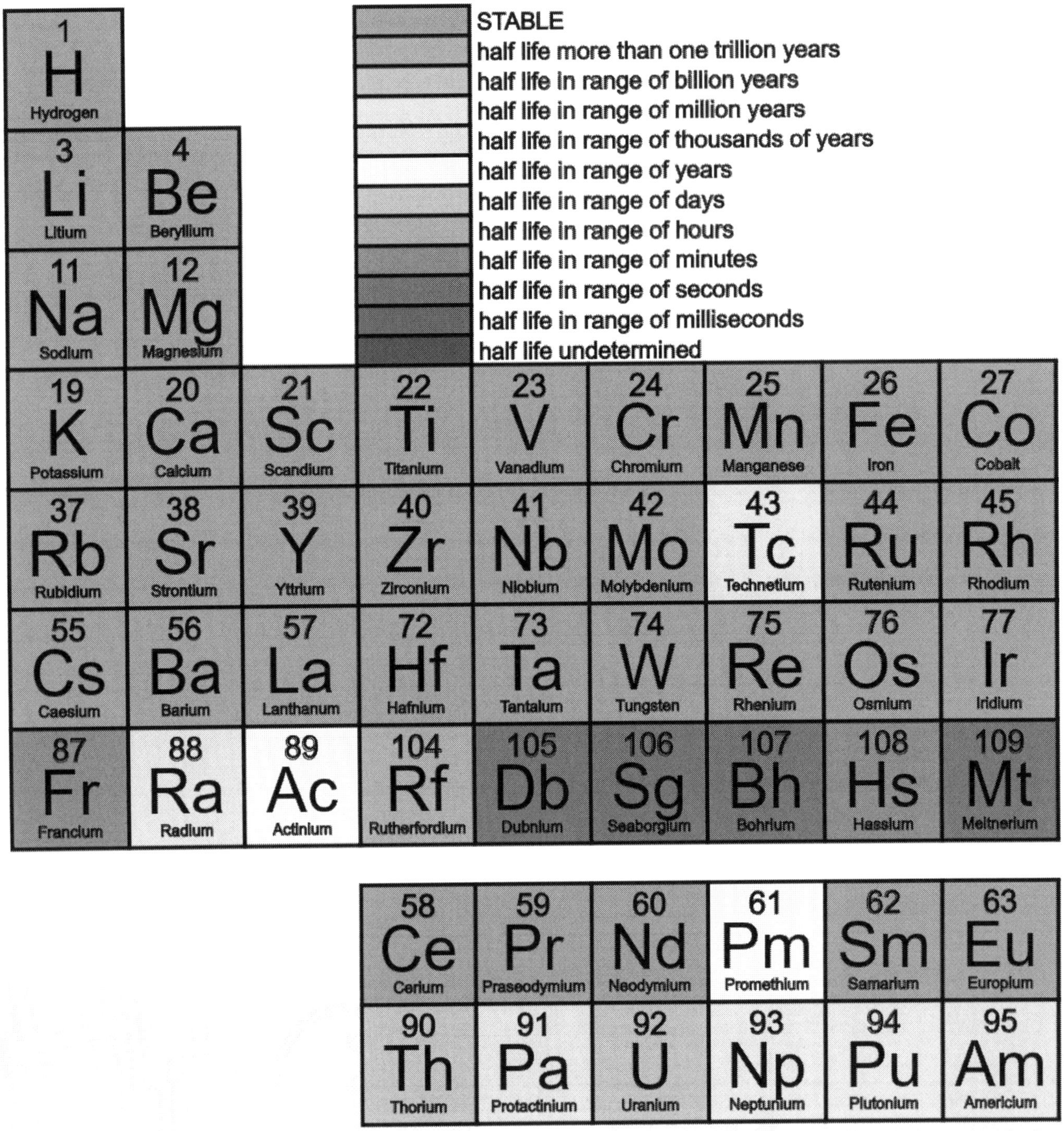

								2 He Helium
			5 B Boron	6 C Carbon	7 N Nitrogen	8 O Oxygen	9 F Fluorine	10 Ne Neon
			13 Al Aluminium	14 Si Silicon	15 P Phosphorous	16 S Sulfur	17 Cl Chlorine	18 Ar Argon
28 Ni Nickel	29 Cu Copper	30 Zn Zinc	31 Ga Gallium	32 Ge Germanium	33 As Arsenic	34 Se Selenium	35 Br Bromine	36 Kr Krypton
46 Pd Palladium	47 Ag Silver	48 Cd Cadmium	49 In Indium	50 Sn Tin	51 Sb Antimony	52 Te Tellurium	53 I Iodine	54 Xe Xenon
78 Pt Platinum	79 Au Gold	80 Hg Mercury	81 Tl Thallium	82 Pb Lead	83 Bi Bismuth	84 Po Polonium	85 At Astatine	86 Rn Radon
110 Ds Darmstadtium	111 Rg Roentgenium	112 UUb Ununbium	113 UUt Ununtrium	114 UUq Ununquadium	115 UUp Ununpentium	116 UUh Ununhexium	117 UUs Ununseptium	118 UUo Ununoctium

64 Gd Gadolinium	65 Tb Terbium	66 Dy Dysprosium	67 Ho Holmium	68 Er Erbium	69 Tm Thulium	70 Yb Ytterbium	71 Lu Lutetium
96 Cm Curium	97 Bk Berkelium	98 Cf Californium	99 Es Einsteinium	100 Fm Fermium	101 Md Mendelevium	102 No Nobelium	103 Lr Lawrencium

Elements and Symbols

Observe the given image of the periodic table and answer the questions that follow.

1. Write the element names for the following symbols.

a. C - ____________

b. H - ____________

c. Al - ____________

d. B - ____________

e. I - ____________

f. Au - ____________

g. Ag - ____________

h. Fe - ____________

i. Pb - ____________

j. He - ____________

2. Write the symbols for the following element names.

a. Argon - ____________

b. Oxygen - ____________

c. Zinc - ____________

d. Uranium - ____________

e. Platinum - ____________

f. Silicon - ____________

g. Lithium - ____________

h. Sulfur - ____________

i. Fluorine - ____________

j. Sodium - ____________

Periodic Table crossword

Complete the crossword by answering the given questions about the periodic table.

Across:

4. Substance that has only one type of atom

7. The most reactive metal

8. How many elements are there in the periodic table?

9. The number of protons in the nucleus

13. Positively charged particles

14. First element

16. What are the rows in the periodic table called?

Down:

1. The most reactive odorless gas

2. Number of rows in the periodic table

3. The basic unit of chemical element

5. Negatively charged particle

6. Number of groups

7. Two or more elements combine to form a ____________ .

10. Who created the periodic table?

11. What are the columns in the periodic table called?

12. A group of atoms bonded together

14. Second element

15. Atomic mass - Proton = ____________

1
2
3
4
5
6
7
8
9
10
11
12
13
14
15
16

Around the World on a Periodic Table

Find elements corresponding to the atomic numbers listed and spell out the places from around the world. Refer to the periodic table to find the elements.

No	Atomic Numbers													Name of the Place
a.	91	19	53	16	73	7								
b.	15	68	92											
c.	6	1	53	11										
d.	53	58	57	60										
e.	10	90	68	57	7	110								
f.	16	91	53	7										
g.	16	53	7	31	84	75								
h.	35	53	73	49										
i.	9	88	7	58										
j.	92	28	52	110	73	52	16	8	9	95	68	53	20	
k.	84	57	60											
l.	18	32	7	22	11									

Add-on: Use a world map to locate the places you have identified.

Element Puns

Identify the element using the clues. Refer to the periodic table of elements, if necessary.

No.	Clues	Elements
1.	What do you do to a wrinkled shirt?	
2.	"Tasting" part of your mouth	
3.	Are extinct	
4.	Mr.Mony's enemy	
5.	What does one say on their deathbed?	
6.	A car or planet	
7.	What do you say to a storm cloud?	
8.	A joint in your leg	
9.	A police officer	
10.	The golden state	
11.	Carbonated drink	
12	A nice person	

Burning of Matter

Let's do a simple activity to prove that oxygen is essential for matter to burn.

What you need:

- Glass jar
- Candle, smaller than the jar
- Plate
- Mittens
- Stopwatch
- Matchstick

What to do:

Step 1 Light the candle on the plate, and keep it on a flat surface.

Step 2 Place the jar inverted over the candle.

Step 3 Start the stopwatch immediately.

Step 4 Wait for the flame to go out. Turn off the stopwatch. Note down the time taken for the flame to go out.

Step 5 Let the jar cool down. Observe the inside of the jar.

Step 6 Wear the mittens and carefully lift the jar and place it quickly on the flat surface.

Step 7 Repeat Steps 1 to 4. Record your observation.

Observation

1. How long does it take for the flame to go out the first time?

2. Does the candle burn the second time?

3. How long does the flame take to go out the second time?

Inference

4. Why do you think this happens?

5. What is produced when the candle burns?

6. What do you think is the explanation for why the candle went out?

Types of Substances

1. Match each diagram with its description. Give two examples for each of the types to complete the table.

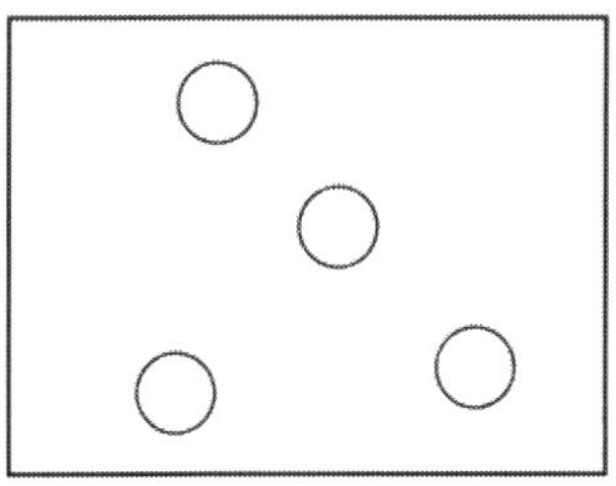

A

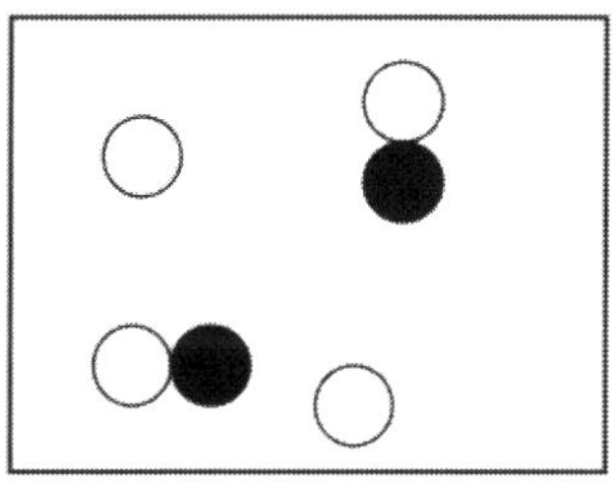

B

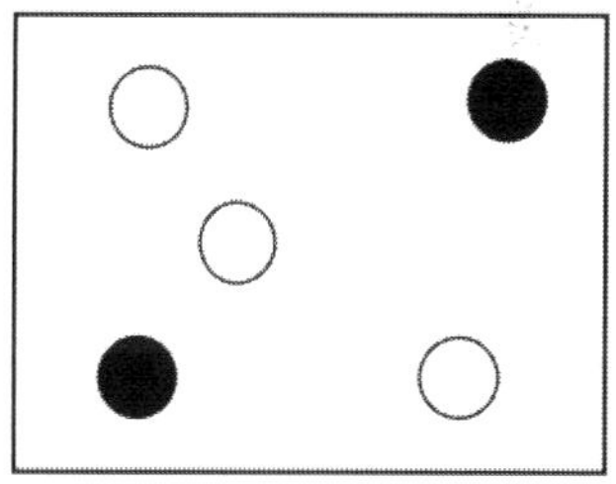

C

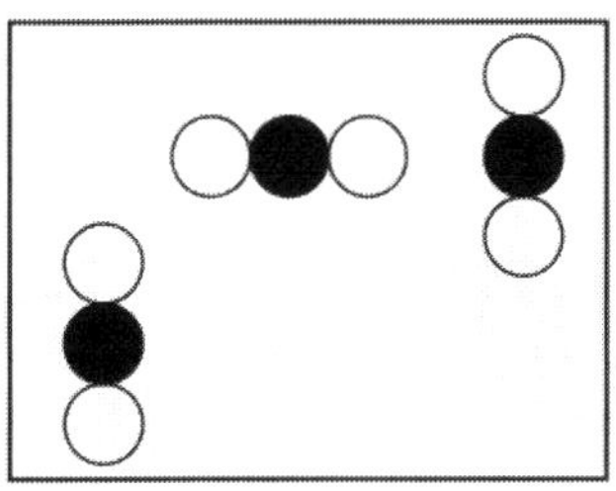

D

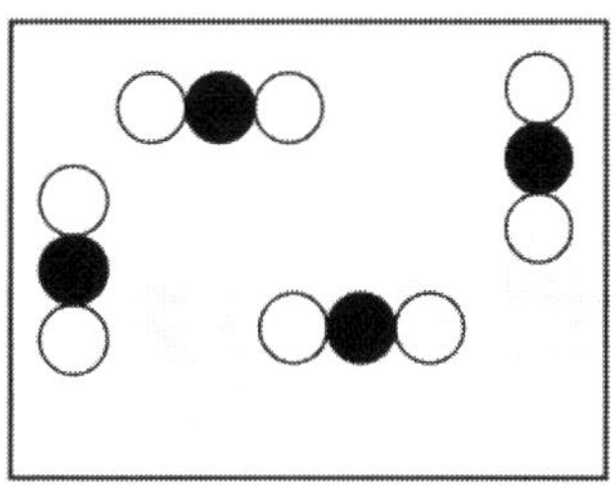

E

Image No.	Description	Examples
	Mixture of a compound and an element	
	Mixture of two elements	
	Pure compound	
	Mixture of two compounds	
	Pure element	

2. Different substances are listed in Column A. Identify their type as elements, compounds or mixtures and write them in Column B. Use Column C for mixtures only. Mention the properties that you could use to separate the mixture.

No.	Column A	Column B	Column C
1	Lemonade		
2	Italian dressing		
3	Salt and pepper		
4	Iron filings and sand		
5	Silver		
6	Soda		
7	Steam		
8	Burrito		
9	Toothpaste		
10	Chicken soup		

Creative Writing

Let's do a simple activity to prove that oxygen is essential for matter to burn.

compound	mixture	element	atom	electron
neutron	proton	substance	properties	metals

Life Sciences

Plants vs Animals

How are plants different from animals? Place a tick mark if a characteristic is applicable.

Characteristics	Plants	Animals
Draw an example and name it		
Eat	☐	☐
Reproduce	☐	☐
Grow	☐	☐
Make food	☐	☐
Need air	☐	☐
Respond to stimuli	☐	☐
Form seeds	☐	☐
Have chlorophyll	☐	☐
Need sunlight	☐	☐
Move on land	☐	☐
Can fly	☐	☐

Plant cell and Animal cell

Students of grade 5 made a model of the plant and animal cells. Mark the parts of the cell in the images of the models provided. Also draw any other organelles you think are missing in the model.

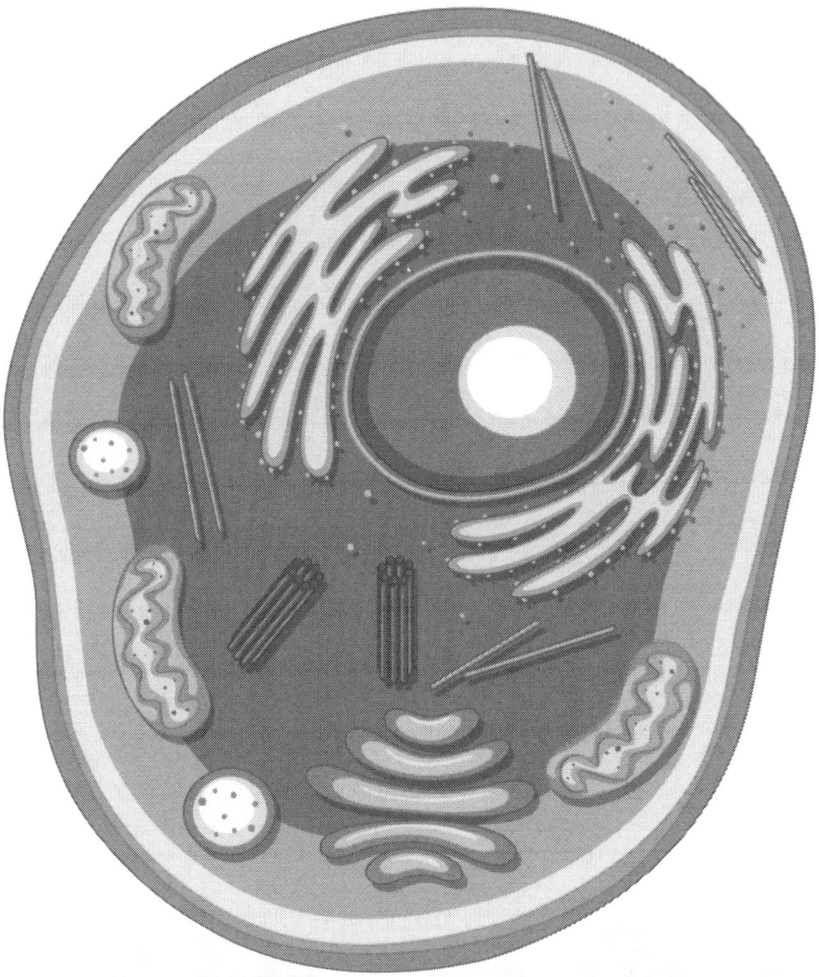

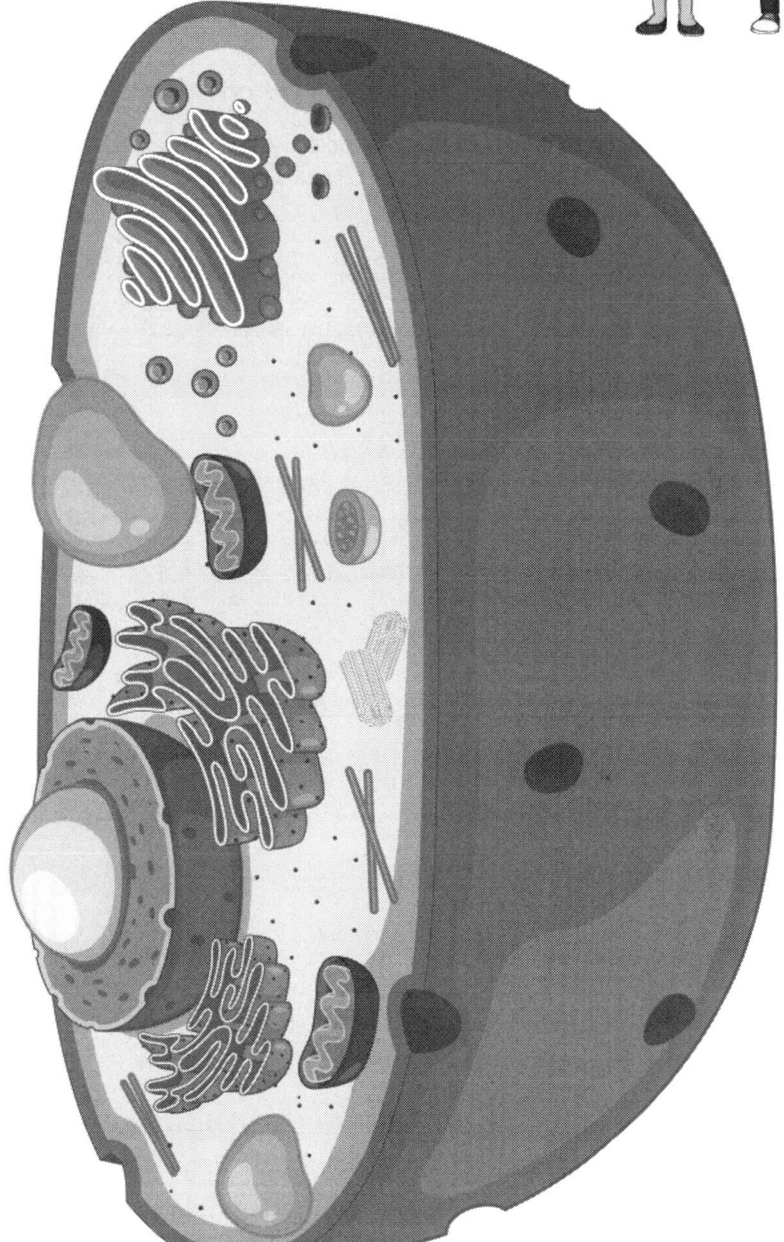

Systems in our Body

Identify the different systems in our body and write the names of two organs involved in that system.

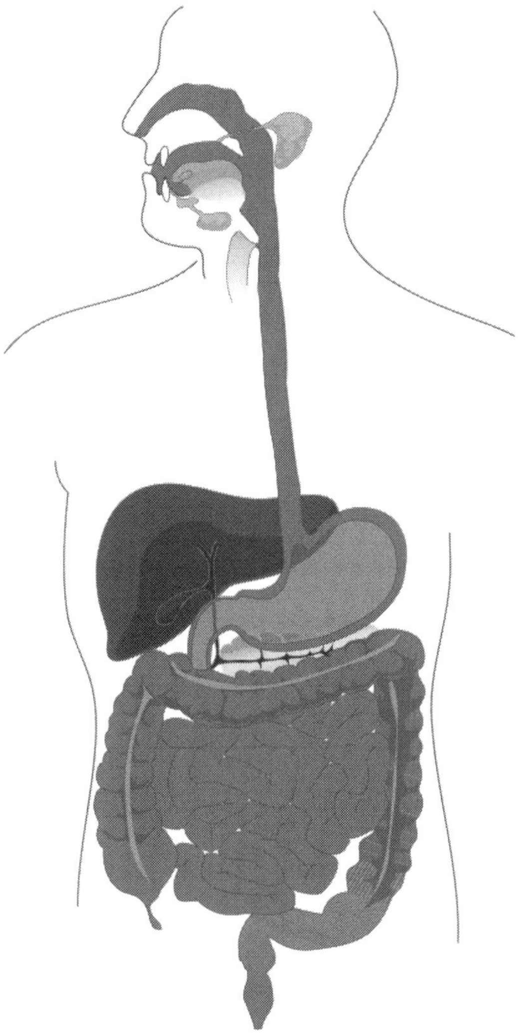

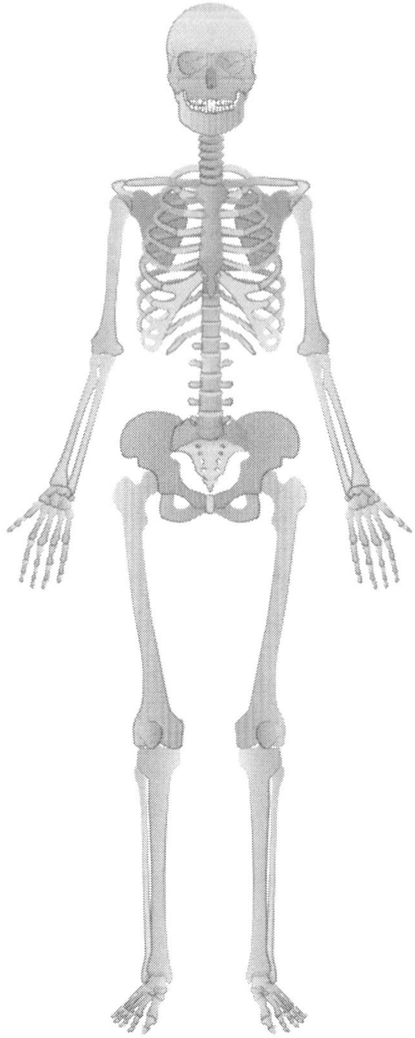

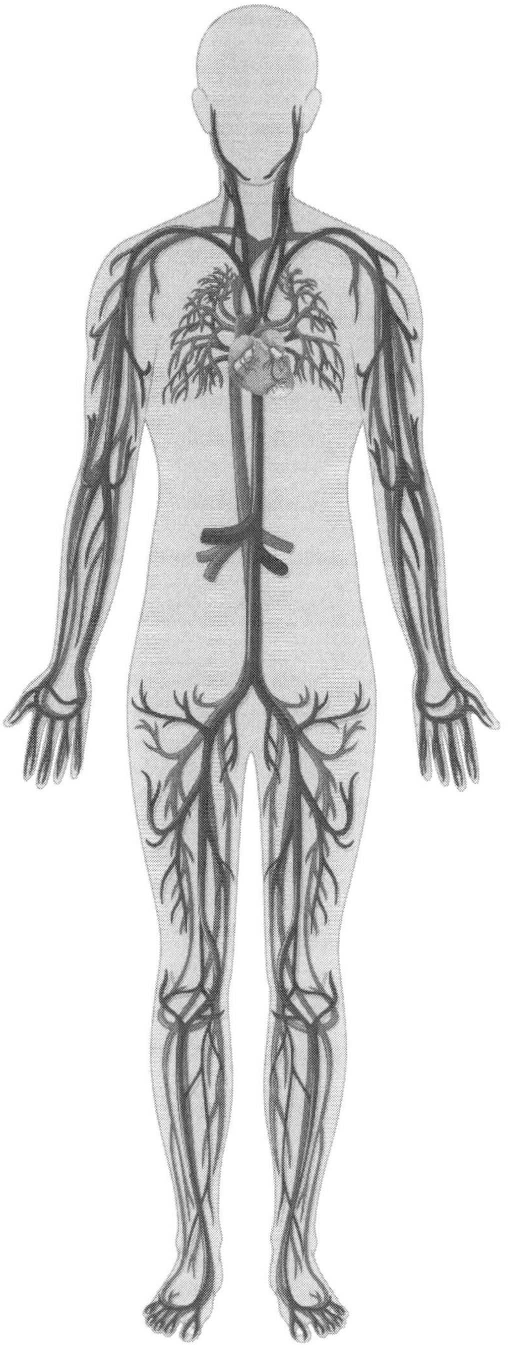

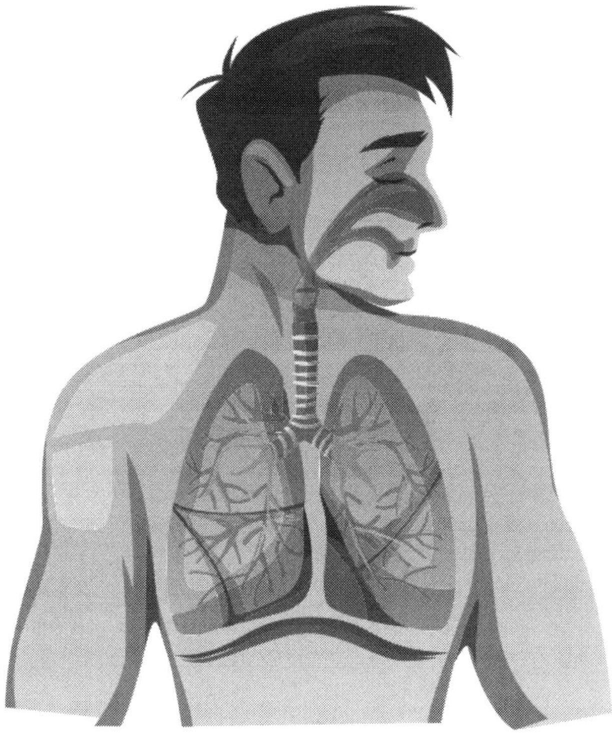

Human Heart

Mark the parts of the human heart. Colour the diagram appropriately.

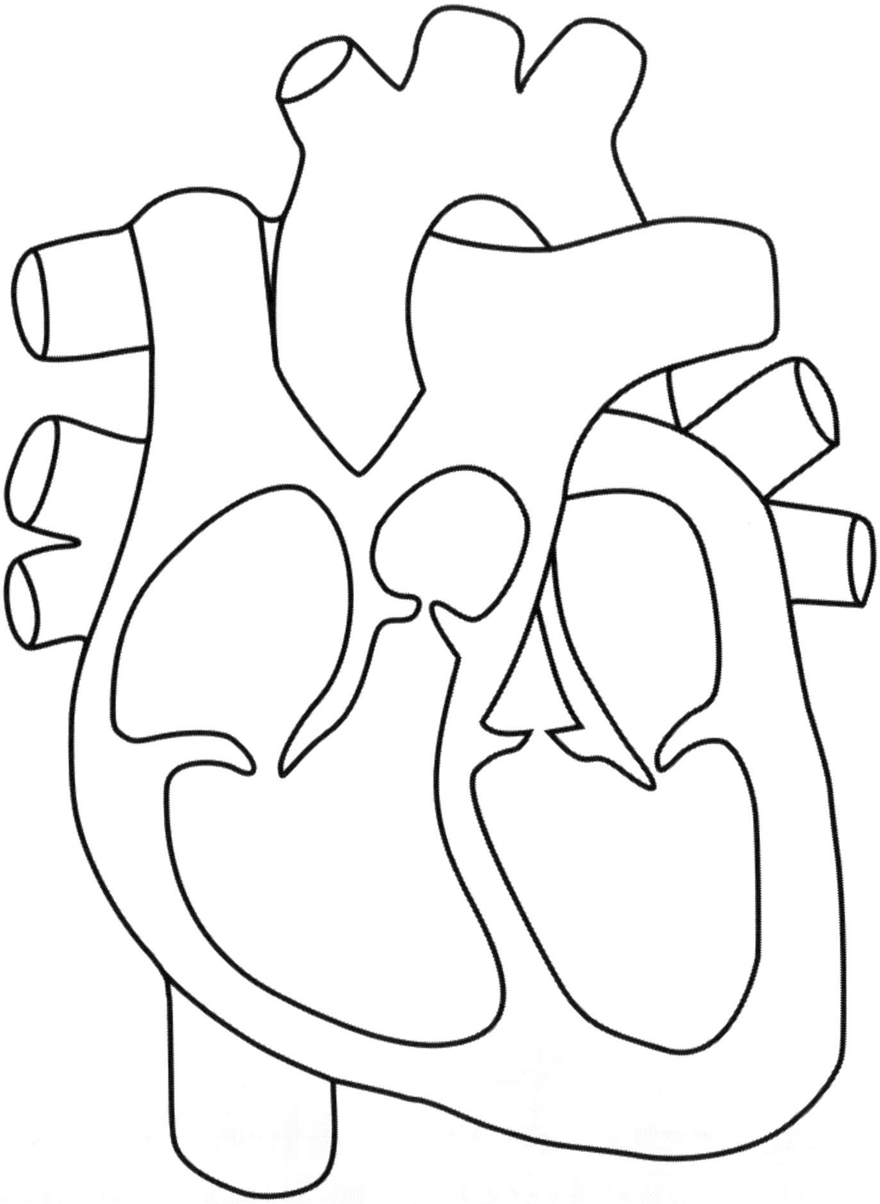

Circulatory System

Color the puzzle piece red if the specified organ is a part of the circulatory system. Color the puzzle piece blue otherwise.

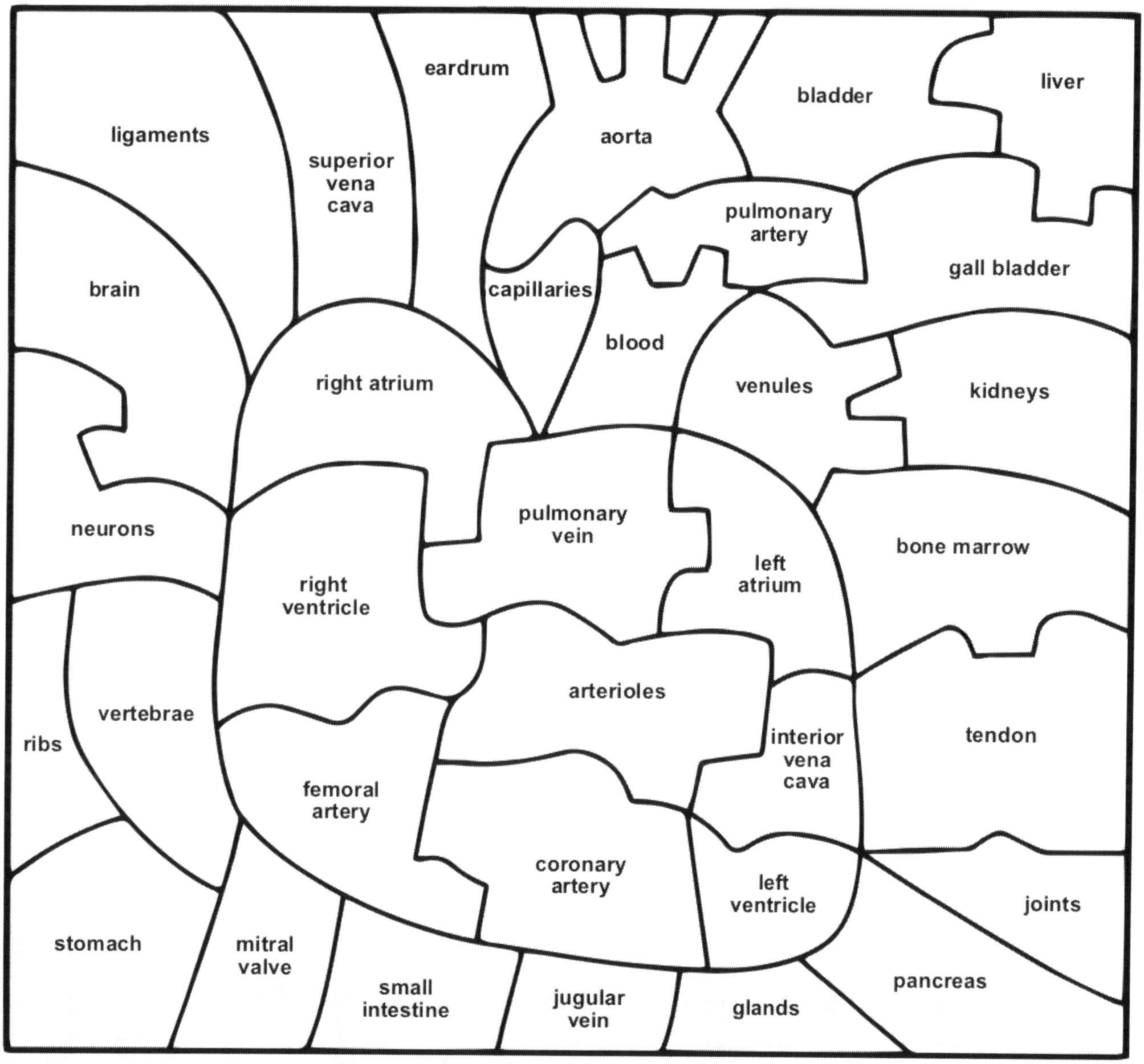

Functions of the Circulatory System

Observe the colouring activity on the circulatory system. Pick any 10 of the organs that are coloured red and list them in the table given below. Write one function of each of these parts in the circulatory system.

No	Name of the organ	Function in the Circulatory System
1		
2		
3		
4		

5		
6		
7		
8		
9		
10		

Steps in Digestion

Match the organs in the digestive system with the listed functions.
Also, identify and name each organ.

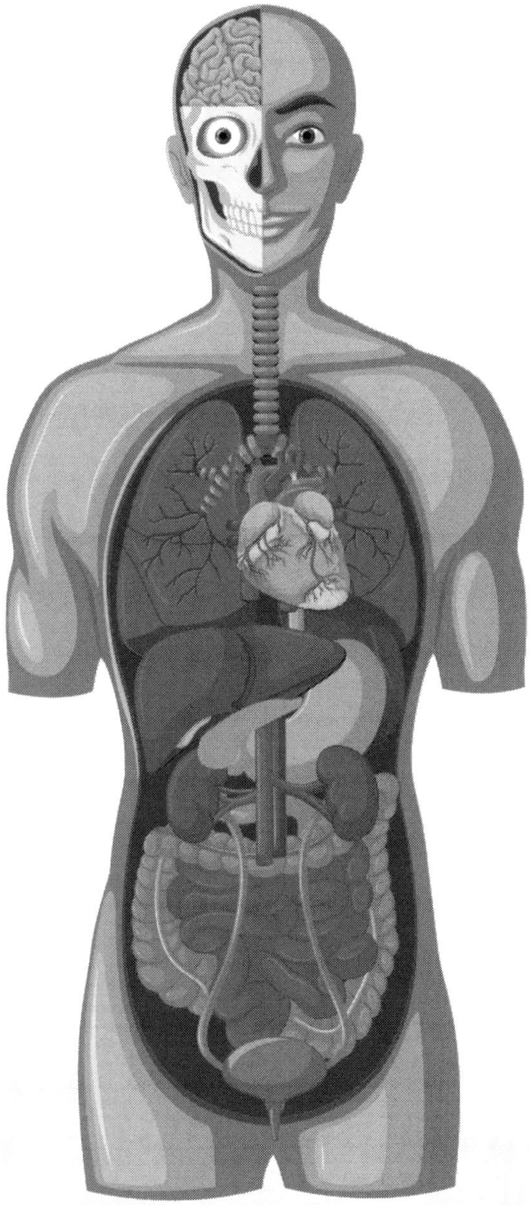

helps chew the food
saliva starts digesting the food
peristalsis pushes the food towards the stomach
secretes the bile
turns food into a soupy liquid
stores bile until needed
digested food mixes with bile and other juices
undigested food and water are processed
solid wastes leave the body

Story Writing

Imagine you are a piece of chocolate in the mouth of a human. Describe your journey through the different organs in the digestive tract. Remember to mention the different feelings and sensations that you will experience through the journey. Draw diagrams wherever necessary.

Urinary System

Label the parts of the excretory system. Also write one function for each of the parts in the table below.

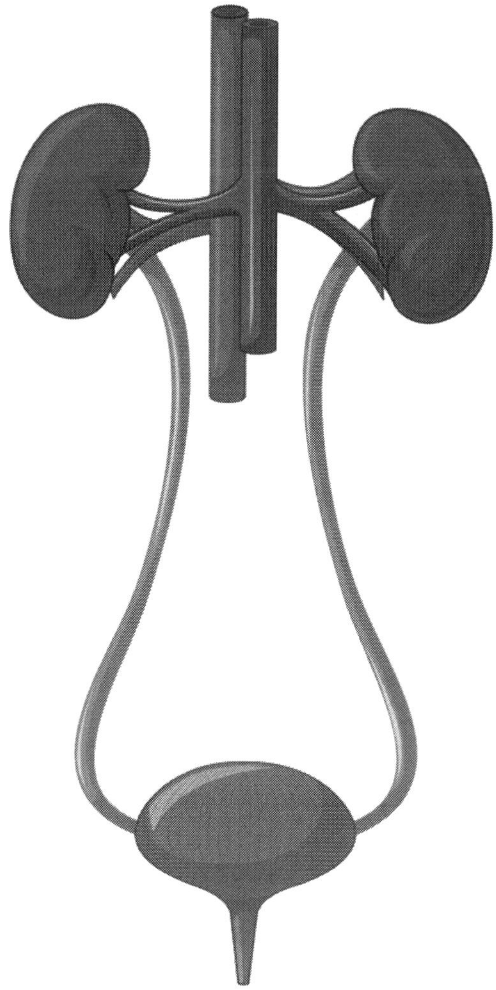

Parts of the urinary system	Functions
Kidneys	
Ureters	
Bladder	
Urethra	

Human Kidney

Label the different parts of the human kidney.

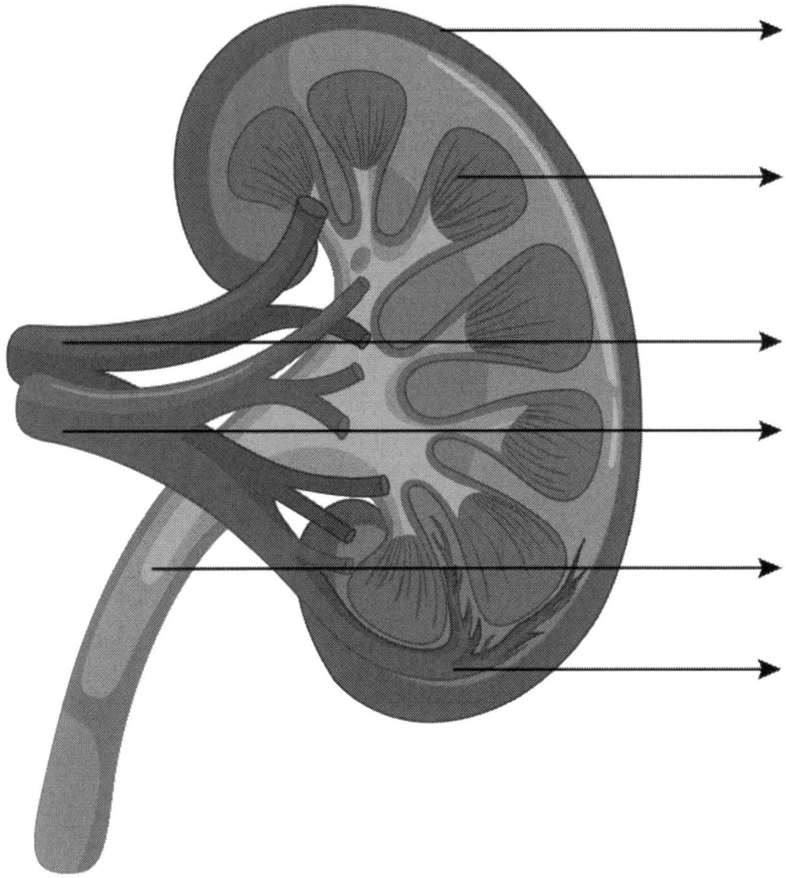

Did You Know?

Each kidney is about the size of a mobile phone. They are bean shaped. The kidney beans are named after the organ as they have a similar shape. A kidney has about a million nephrons. The right kidney is slightly smaller than the left kidney to make room for the liver. Humans do not need both the kidneys to survive. One can live a healthy life with just one kidney.

DIY Lung Model

Would you want to see your lungs in action by making a model?

What you need:

- Single-use, plastic 2-litre bottle
- 2 balloons
- 2 elastic bands
- Modelling clay
- A pair of scissors
- Masking tape
- Drinking straw

What to do:

Step 1 With the help of an adult, cut the plastic bottle in half.

Step 2 Put a straw into a balloon and seat the opening tight with masking tape. Check sealing by blowing air into the free end of the straw. If the balloon doesn't puff up, check and seal with masking tape better.

Step 3 Insert this straw-balloon setup into the mouth of the cut plastic bottle.

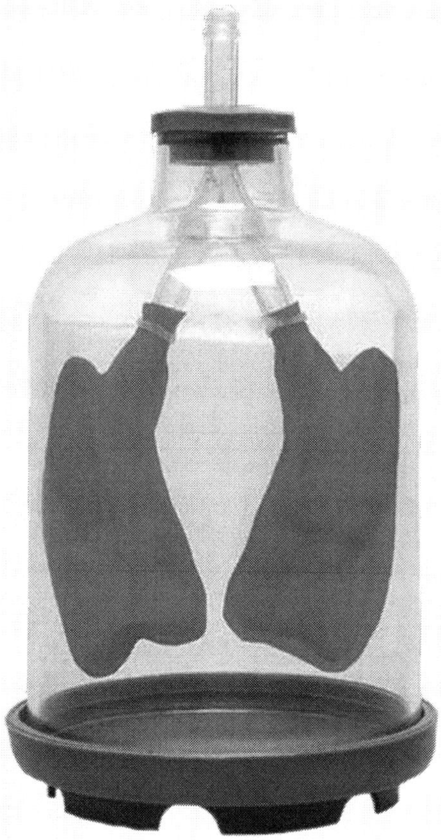

Step 4 Secure the straw at the mouth of the bottle using the modelling clay.

Step 5 Tie a knot at the end of the other balloon.

Step 6 Cut the balloon in half.

Step 7 Use the piece of the balloon with the knot, stretch, open the other end over the bottom of the plastic bottle.

Step 8 Gently pull down the balloon by the knot. This causes air to flow into the bottle and then into the balloon causing it to expand like a lung when you inhale.

Step 9 Release the balloon to see the air being expelled from the balloon, like when it happens when you exhale.

Respiratory System

1. Label the parts of the respiratory system and answer the questions that follow.

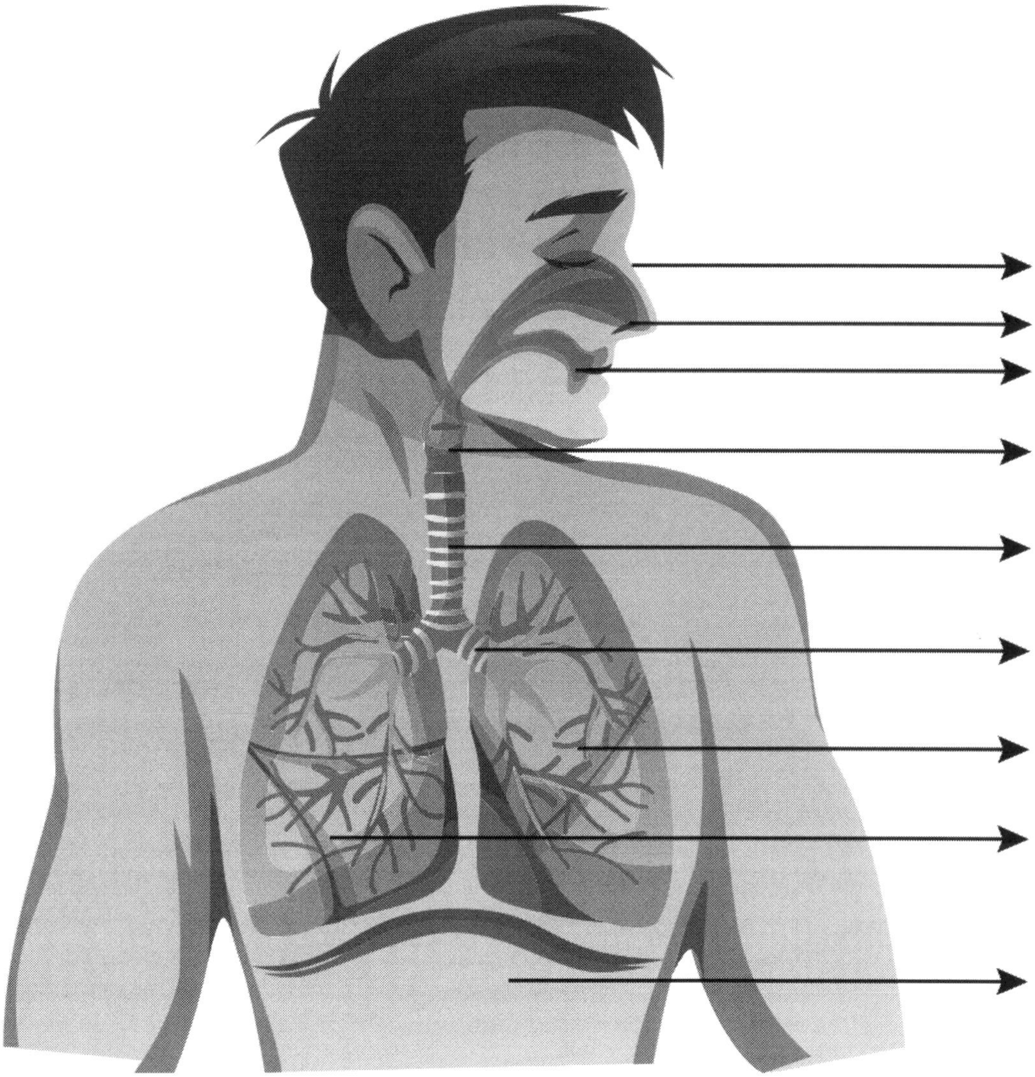

2. Draw bronchioles and alveoli in the right lung.

3. Fill in the blanks.

- The ____________ is where exchange of gases takes place.
- The trachea are supported by the c-shaped rings of ____________.
- Air enters the lungs when we ____________, and exits when we ____________.
- The larynx is also known as the ____________.
- The ____________ is a sheet of muscle that divides the thorax from the abdomen.
- ____________ is the name of the fluid that lubricates the lungs.
- The ____________ is also called the windpipe.

Exchange of Gases

Observe the arrows and label the diagram appropriately. Use the word grid given below.

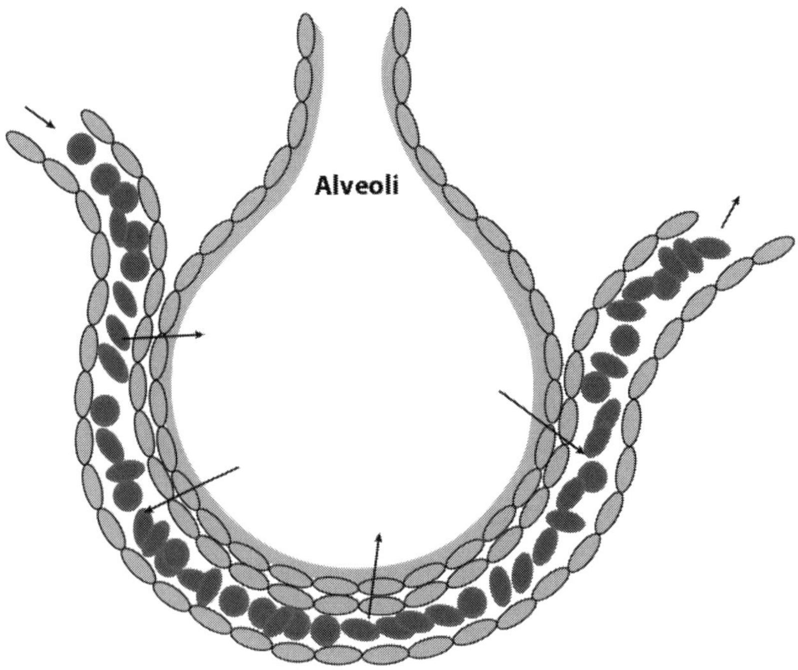

pure blood	impure blood	oxygen	carbon dioxide

Photosynthesis and Cellular Respiration

Answer the questions about photosynthesis and cellular respiration.

1. What is the general chemical equation for photosynthesis?

____

____

____

2. What is the general chemical equation for cellular respiration?

____

____

____

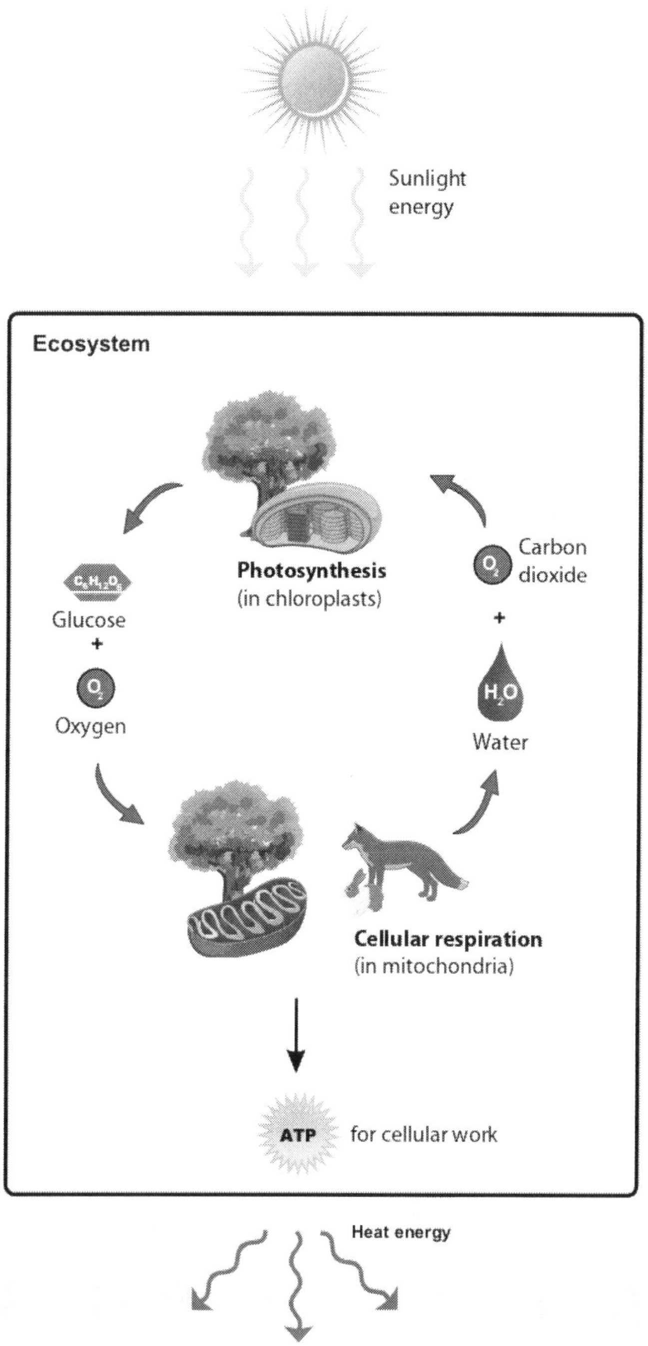

Observe the picture and describe the relationship between photosynthesis and cellular respiration.

____

____

____

____

____

____

Earth Sciences

Pop Quiz

Answer the following questions on the Solar System.

1. What is the seventh planet from the Sun?

____

2. Which planet has the largest ring system?

____

3. Which is the coldest planet?

____

4. Which is the hottest planet?

____

5. Which planet supports life?

____

6. Which planet is so light, that it could float on water?

____

7. What is called the 'red planet'?

8. What is the brightest object in the sky?

9. Which is the largest planet?

10. Which planet has no atmosphere?

11. What is called the planet with a great red spot?

12. Which planet is dominated by volcanoes?

13. Which planet has the highest mountain?

14. Which is the only star in our solar system?

15. Which planet has the most number of moons?

Fact or opinion?

Read each statement and decide if it is a fact or opinion.

1. Saturn's rings are beautiful.	
2. The planet closest to the Sun is Mercury.	
3. A galaxy is a group of stars.	
4. All students should see a meteor shower.	
5. Venus is the second planet from the Sun.	
6. Comets orbit the Sun	
7. The moon is beautiful.	
8. The Sun is a yellow star.	
9. Everyone likes to look at pictures from space.	
10. A comet's tail always faces away from the Sun.	
11. Mars has two moons.	
12. I would like to visit Mars.	
13. Asteroids orbit between Mars and Jupiter.	
14. The Moon is beautiful.	
15. Everyone should study astronomy.	

Planet Report

Answer the following questions to complete a report on your favourite planet. Use the internet, if necessary.

Draw and colour a picture of the planet here.

1. Name of the planet:

2. The planet was named after:

3. Number of moons the planet has:

4. Number of Earth days the planet takes to orbit the Sun:

5. Number of Earth days the planet takes to rotate about its own axis:

6. Write 4 fun facts about the planet.

Water cycle

Define the different terms regarding the water cycle and label them in the diagram given below.

1. Accumulation

____

____

2. Condensation

____

____

3. Evaporation

____

____

4. Precipitation

____

____

5. Runoff

____

____

6. Transpiration

____

____

Layers of the Atmosphere

Colour and label the different layers of the atmosphere with a colour key.

- Exosphere
- Troposphere
- Thermosphere
- Stratosphere
- Mesosphere

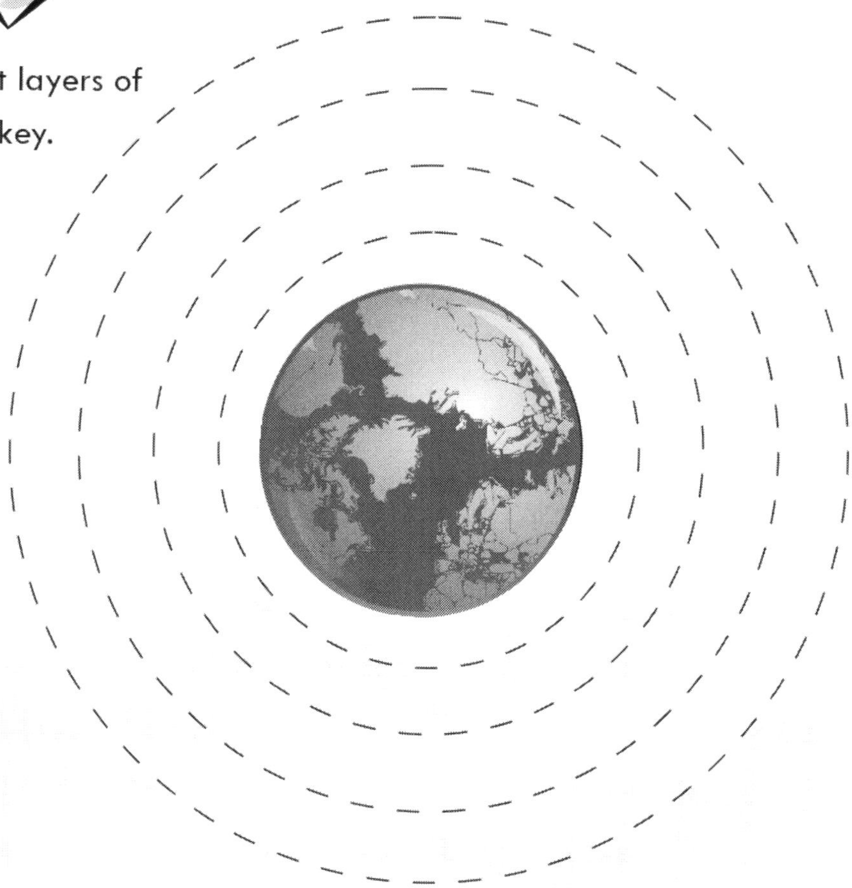

Sea Breeze Vs Land Breeze

In each of the drawings:

1. label the areas of high pressure and low pressure.
2. draw the movement of wind using arrows.
3. answer the question that follows.

1. Explain how sea breezes form.

____

____

____

____

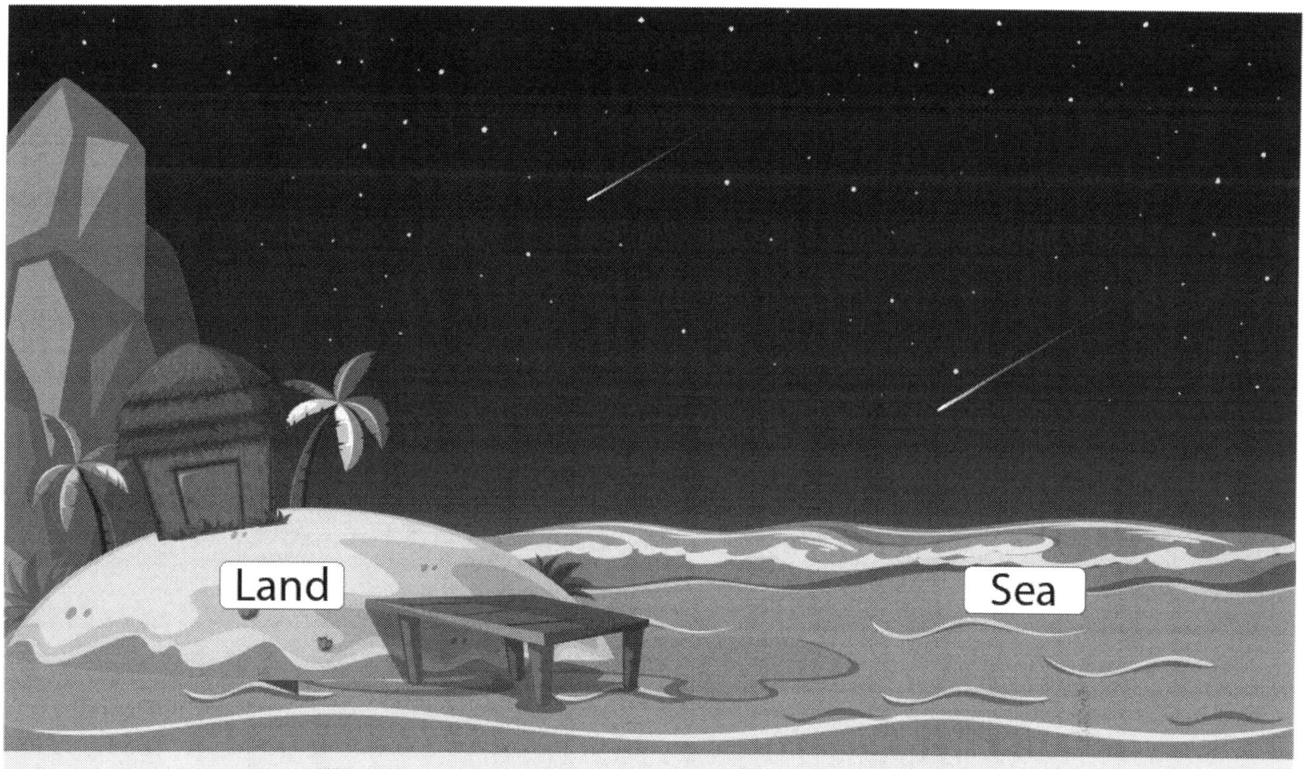

2. Explain how land breezes form.

3. In the diagrams, what are the circular motions of air called?

4. What causes the circular motion of air to occur?

5. Why is the warmer air rising while the cold air sinks?

6. Why is the land warmer during the day and cooler during the night, compared to the sea?

7. During the daytime, would you face towards the land or the sea to feel a breeze blowing on your face?

Weather Vocabulary

Write one word that perfectly suits each of the descriptions given below.

Note: Some answers may be two words.

1. A destructive funnel cloud moving over a narrow path of land	
2. A strong, narrow band of wind with very fast-moving air that moves across the United States from west to east	
3. A sudden discharge or spark of electricity between two clouds or between a cloud and the ground	
4. Fluffy, white clouds resembling balls of cotton that are usually associated with fair weather	
5. What forms when condensed water from rivers, lakes, moist plants, or soil floats in the air near the ground	
6. Feathery, fibrous clouds made of ice crystals that are usually associated with fair weather	
7. A form of air pollution appearing as a thick, brownish haze that is formed when carbon monoxide and other gases react in sunlight	

8. A severe rainstorm that occurs when a cold front meets a warm front and cumulonimbus clouds form in the air	
9. The heating of the earth's atmosphere caused by a buildup of carbon dioxide when humans burn excessive amounts of fossil fuels	
10. A boundary that is formed when two air masses with different properties meet	
11. The amount of moisture in the air compared to the amount it can hold at a given temperature	
12. Large tropical storms with high winds and heavy rain that can cause great damage as they move over land	
13. A severe winter snowstorm that can cause near zero visibility with extremely low temperatures and high winds	
14. The average of all weather conditions of an area over a long period of time	
15. Thick, layered gray clouds that usually cover the sky in an even blanket and are usually associated with rain and snow	

ocean currents

Answer the questions about ocean currents given below.

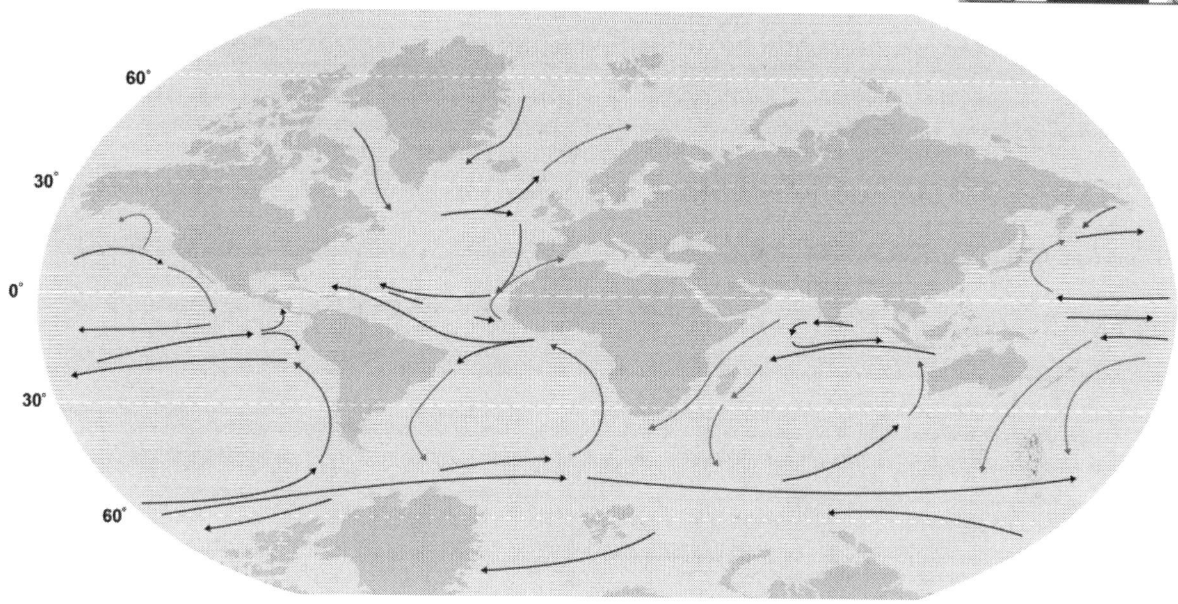

1. Label the ocean currents in the given map of the world. Also complete the table given below.

Name of the current	Characteristic temperature of the water transported by the current (cold current/warm current)
Antarctic circumpolar current	
Benguela current	
Brazil current	
California current	
Canary current	
East Australian current	
Gulf stream	
Kurushio current	
Peru current	

2. Observe the ocean currents on the map. They generally travel in clockwise or counterclockwise direction.

a. The direction of ocean currents in the Northern Hemisphere is ____________.

b. The direction of ocean currents in the Southern Hemisphere is ____________.

3. Look at the pattern of ocean currents in the Northern and Southern hemispheres and describe the effect the currents have on the temperature.

a. The West coasts generally have ____________ (warm/cold) ocean currents.

b. The East coasts generally have ____________ (warm/cold) ocean currents.

c. The climates on the West coast will generally be ____________ (warmer/colder) than it is inland.

d. The climates on the East coast will generally be ____________ (warmer/colder) than it is inland.

Ocean Currents and Weather

Read the passage about ocean currents. Answer the questions that follow.

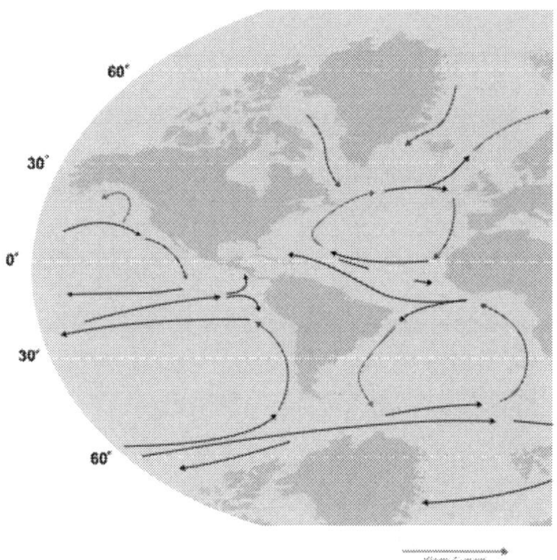

Ocean currents move like the river, except that they flow through the ocean in different directions.

Currents are caused by plate tectonics, earthquake, wind, and gravity. Some currents are warm, while some are cold.

Ocean currents affect the weather on land. Warm currents, like the Gulf Stream, bring warm air, while the cold currents bring cold air.

These currents help shape beaches and coastlines. They also help cause severe weather conditions such as hurricanes and typhoons.

1. How did the North Equatorial and South Equatorial currents get their names?

2. What warm current flows through the east coast of the United States?

3. What causes the ocean currents?

4. Write the names of the currents that you think affect the United States' weather the most.

5. How do warm currents affect weather?

Freshwater Vs Saltwater

Sort the given water bodies based on their characteristics.

Freshwater	Both	Saltwater

Seas	Ponds	Arctic ocean	Stream	Southern ocean
River	Indian ocean	Lake	Pacific ocean	Glacier
Has currents	Natural Resource	Atlantic ocean	Most of Earth's water	Groundwater

Distribution of Water on Earth

Observe the given pie chart and answer the questions that follow.

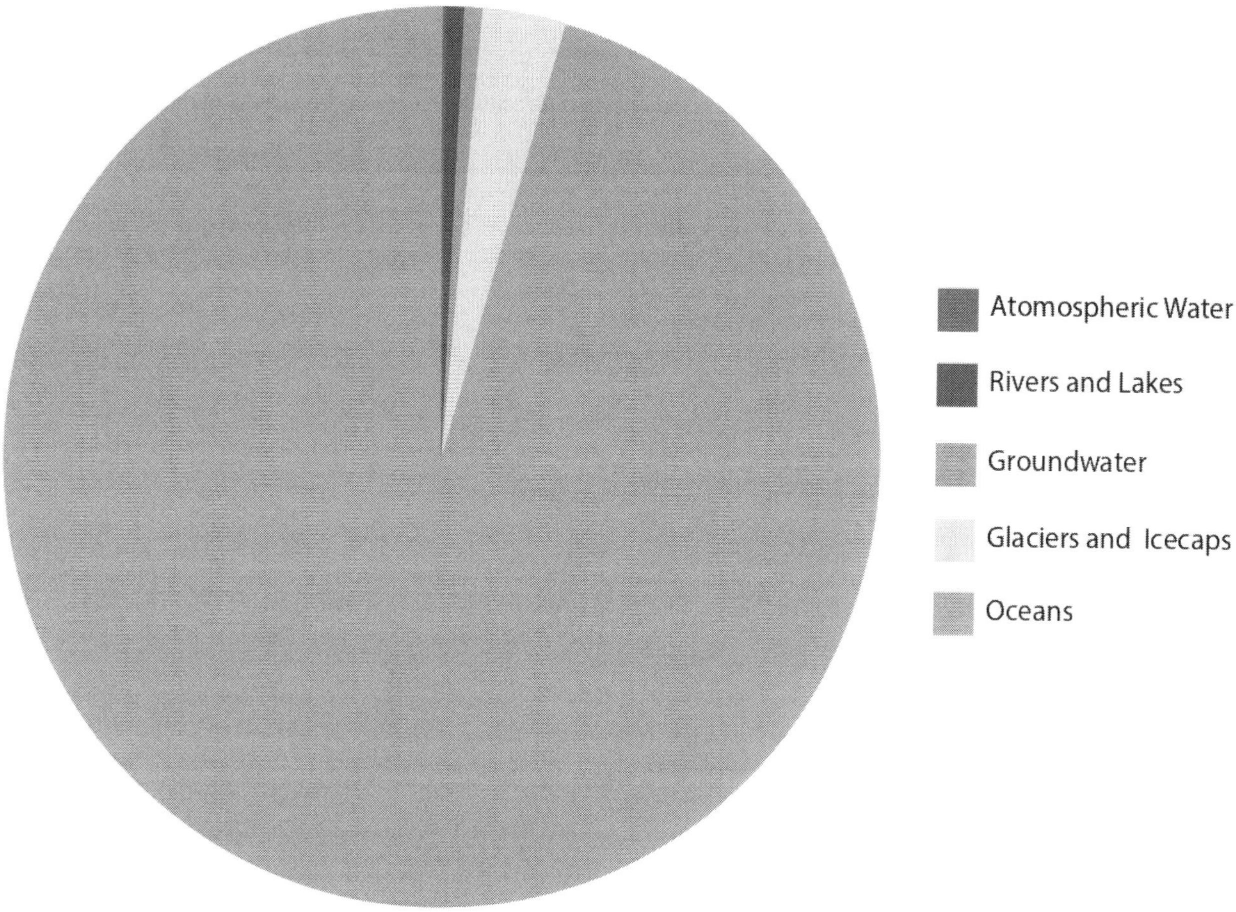

1. What are the different sources of water included in the pie chart?

2. Where is most of Earth's water present?

3. Where is the least of Earth's water present?

4. What do you think is the water that is found in abundance in the Earth's crust - freshwater or saltwater?

5. Why do you think there is more saltwater than freshwater on Earth?

6. Where is more of Earth's freshwater locked up in?

Water Consumption Survey

Talk to 5 adults in your neighbourhood and get them to answer the following survey on domestic water consumption.

Participant 1

Name: ______________________________

1. Does your family use a dishwasher?
a. Yes
b. No

2. Does your family use drinkable water for watering the garden?
a. Yes
b. No
c. No garden

3. Does your family use drinkable water for the swimming pool?
a. Yes
b. No
c. No pool

4. Do you usually leave the tap water running while brushing your teeth?
a. Yes
b. Sometimes
c. Never

5. Do you usually drink tap water or bottled water?
a. Only tap water
b. Only bottled water
c. Both

6. How often do you have a bath?
a. Everyday
b. 3-5 times a week
c. Less than that

7. Do you use a shower or fill a tub with water?
a. Shower
b. Tub
c. Both

8. Have you ever had any medical problems in your house because of the quality of water?
a. Yes
b. No
c. I don't know

9. Do you believe that increased water consumption is an environmental issue?
a. Yes
b. No
c. I don't care

10. Do you run the washing machine and dishwasher only if they are fully loaded or even when they are half-empty?
a. Only when full
b. Even when half-empty
c. Doesn't matter

11. Does any tap in the house doesn't close well and keep dripping?
a. Yes, 1
b. More than 1
c. No leaky taps

12. What is the approximate quantity of water you consume in your house according to the water company?
a. Less than 20 m^3
b. 20 - 40 m^3
c. More than 40 m^3

Participant 2

Name: ______________________________

1. Does your family use a dishwasher?

a. Yes

b. No

2. Does your family use drinkable water for watering the garden?

a. Yes

b. No

c. No garden

3. Does your family use drinkable water for the swimming pool?

a. Yes

b. No

c. No pool

4. Do you usually leave the tap water running while brushing your teeth?

a. Yes

b. Sometimes

c. Never

5. Do you usually drink tap water or bottled water?

a. Only tap water

b. Only bottled water

c. Both

6. How often do you have a bath?

a. Everyday

b. 3-5 times a week

c. Less than that

7. Do you use a shower or fill a tub with water?

a. Shower

b. Tub

c. Both

8. Have you ever had any medical problems in your house because of the quality of water?

a. Yes

b. No

c. I don't know

9. Do you believe that increased water consumption is an environmental issue?

a. Yes

b. No

c. I don't care

10. Do you run the washing machine and dishwasher only if they are fully loaded or even when they are half-empty?

a. Only when full

b. Even when half-empty

c. Doesn't matter

11. Does any tap in the house doesn't close well and keep dripping?

a. Yes, 1

b. More than 1

c. No leaky taps

12. What is the approximate quantity of water you consume in your house according to the water company?

a. Less than 20 m^3

b. 20 - 40 m^3

c. More than 40 m^3

Participant 3

Name: ________________________________

1. Does your family use a dishwasher?

a. Yes

b. No

2. Does your family use drinkable water for watering the garden?

a. Yes

b. No

c. No garden

3. Does your family use drinkable water for the swimming pool?

a. Yes

b. No

c. No pool

4. Do you usually leave the tap water running while brushing your teeth?

a. Yes

b. Sometimes

c. Never

5. Do you usually drink tap water or bottled water?

a. Only tap water

b. Only bottled water

c. Both

6. How often do you have a bath?

a. Everyday

b. 3-5 times a week

c. Less than that

7. Do you use a shower or fill a tub with water?

a. Shower

b. Tub

c. Both

8. Have you ever had any medical problems in your house because of the quality of water?

a. Yes

b. No

c. I don't know

9. Do you believe that increased water consumption is an environmental issue?

a. Yes

b. No

c. I don't care

10. Do you run the washing machine and dishwasher only if they are fully loaded or even when they are half-empty?

a. Only when full

b. Even when half-empty

c. Doesn't matter

11. Does any tap in the house doesn't close well and keep dripping?

a. Yes, 1

b. More than 1

c. No leaky taps

12. What is the approximate quantity of water you consume in your house according to the water company?

a. Less than 20 m^3

b. 20 - 40 m^3

c. More than 40 m^3

Participant 4

Name: ______________________________

1. Does your family use a dishwasher?

a. Yes

b. No

2. Does your family use drinkable water for watering the garden?

a. Yes

b. No

c. No garden

3. Does your family use drinkable water for the swimming pool?

a. Yes

b. No

c. No pool

4. Do you usually leave the tap water running while brushing your teeth?

a. Yes

b. Sometimes

c. Never

5. Do you usually drink tap water or bottled water?

a. Only tap water

b. Only bottled water

c. Both

6. How often do you have a bath?

a. Everyday

b. 3-5 times a week

c. Less than that

7. Do you use a shower or fill a tub with water?

a. Shower

b. Tub

c. Both

8. Have you ever had any medical problems in your house because of the quality of water?

a. Yes

b. No

c. I don't know

9. Do you believe that increased water consumption is an environmental issue?

a. Yes

b. No

c. I don't care

10. Do you run the washing machine and dishwasher only if they are fully loaded or even when they are half-empty?

a. Only when full

b. Even when half-empty

c. Doesn't matter

11. Does any tap in the house doesn't close well and keep dripping?

a. Yes, 1

b. More than 1

c. No leaky taps

12. What is the approximate quantity of water you consume in your house according to the water company?

a. Less than 20 m^3

b. 20 - 40 m^3

c. More than 40 m^3

Participant 5

Name: ______________________________

1. Does your family use a dishwasher?

a. Yes

b. No

2. Does your family use drinkable water for watering the garden?

a. Yes

b. No

c. No garden

3. Does your family use drinkable water for the swimming pool?

a. Yes

b. No

c. No pool

4. Do you usually leave the tap water running while brushing your teeth?

a. Yes

b. Sometimes

c. Never

5. Do you usually drink tap water or bottled water?

a. Only tap water

b. Only bottled water

c. Both

6. How often do you have a bath?

a. Everyday

b. 3-5 times a week

c. Less than that

7. Do you use a shower or fill a tub with water?

a. Shower

b. Tub

c. Both

8. Have you ever had any medical problems in your house because of the quality of water?

a. Yes

b. No

c. I don't know

9. Do you believe that increased water consumption is an environmental issue?

a. Yes

b. No

c. I don't care

10. Do you run the washing machine and dishwasher only if they are fully loaded or even when they are half-empty?

a. Only when full

b. Even when half-empty

c. Doesn't matter

11. Does any tap in the house doesn't close well and keep dripping?

a. Yes, 1

b. More than 1

c. No leaky taps

12. What is the approximate quantity of water you consume in your house according to the water company?

a. Less than 20 m^3

b. 20 - 40 m^3

c. More than 40 m^3

Answers

English Answer Key

Verbs

Present or Past?

1. come, came
2. eat, ate
3. hear, heard
4. went, go
5. give, gave
6. took, take

Complete Me!

1. lost
2. spoke
3. gave
4. told
5. drank

Verb Tenses

Conjugate

1. traveling
2. eating
3. has
4. visits
5. jogs
6. played
7. would have completed
8. plan/are planning/will be planning
9. would have reached
10. slept
11. crash
12. rolled

Fix Me!

1. Maria has been attending Spanish class since last week.
2. When the oven beeps, she will take the cake out.
3. I was wondering if you cooked last night.
4. After school, my friends went to the park, but I went home.
5. Every morning, we ride to school by bus.

Make Sentences

Answers will vary

Fill in the Blanks

1. be joining us
2. will be playing
3. won't be able to
4. will submit
5. will go

Form Future Tense

1. When he comes next year, we will have moved to a new house.
2. By the time you buy lunch, I will have reached home.
3. In five years, I will have finished my higher studies.
4. She will have completed the book before she retires.
5. My mother will have cooked breakfast before you get back.
6. Please, call back later. He will have returned by two o'clock.
7. I will have read by tomorrow morning.
8. Jason said that he will have completed the assignment by Monday morning.

Complete Me

1. will be selling
2. will be going
3. will be coming
4. will leave
5. will rain

Conjunctions

This or That?

1. until
2. or
3. Unless
4. before
5. but
6. Though
7. and
8. both
9. Although
10. if

Combine Sentences

1. The boys were standing to the right of the stage and the girls to the left.

2. I don't have my phone because I left it in my other handbag.

3. The teacher kept giving examples until the students understood the theory.

4. My dog is scared of noise, so I take him to the park in the morning when it is quiet.

5. I was feeling hungry, but mother asked me to wait for an hour.

Correlative Conjunctions

Identify the Conjunctions

1. both, and
2. not only, but also
3. either, or
4. neither, nor

Complete Me!

1. not only, but also
2. either, or
3. not only, but also
4. neither, nor / either, or
5. either, or
6. neither, nor / either, or

Spelling

Spelling Staircase

1. day
2. Desk
3. Diary
4. Decide
5. Dazzled
6. Decorate

Word Game

Answers will vary

Find the Missing Letters

1. Pancakes
2. Spatula
3. Frying Pan
4. Butter
5. Recipe
6. Ingredient
7. Cooking

This or That?

Calendar; Beginning; Exclusive; Remaining; Portrait; Endeavor; Advance; Happening

Words within Words

camp, vision, drop, display, follow, rough, direct, exam, hole, birth

Context Clues

Decode the Clues

1. a. strange
2. c. rejected
3. b. rude
4. c. motivation
5. a. not moving

Affixes and Root Words

Make New Words

semicircle; semifinal;
semiarid; semirural;
semipro; semisolid;
semidomes; semiskilled

Add Suffixes

comfort + able, sad + ness, diction + ary,
victor + y, hope + ful, live + ly, punctuate + tion,
light + ing; scholar + ship; work + ed

Split the Words

1. mater (root word) + ial (suffix)
2. jud(root word) +ge (suffix)
3. port (root word) + able (suffix)
4. ad (prefix) + here (root)
5. love (root word) + ly (suffix)
6. trans (prefix) + mit (root)
7. mis (prefix) + spell (root)
8. sub (prefix) + way (root)

Reference Materials

Find Me!

Answers may vary

Sort Me

Activity – Action, Movement, Bustle, Hobby

Maximum – Full, Greatest, Top, Utmost

Table – Desk, Chart, List, Bench

Treatment – Cure, Healing, Handling, Care

Synonyms and Antonyms

Match the Synonyms

subside - abate

benevolent - charitable

diligence - persistence

sufficient - adequate

exploit - utilize

paramount - supreme

Complete Me!

1. difficult
2. fast
3. healthy
4. create
5. scared

Homographs

Find the Meaning

Bear – an animal; to tolerate

Minute – tiny; unit of time

Close–near; shut

Bank – financial institute; the land area near a waterbody

Bass – a deep-toned sound; a type of fish

Capital – the headquarters of a country or a state; money

Desert – a dry land area; to abandon

Make Sentences

Answer may vary

Transitional Words

Connect Sentences

Answer may vary

Connect the Events

a. first b. next c. before d. after e. When

Who Will Bell the Cat?

True or False

1. False
2. False
3. True
4. False
5. True

Read and Answer

1. The baker sold bread, pies, muffins, bagels, bread rolls, buns, and more.
2. The mice were attracted by the smell of the freshly-baked food.
3. When the baker saw all his cakes destroyed by the mice, he decided to get a cat.
4. The cat walked so softly that the mice were not able to hear him coming. So, they decided to bell the cat.

Describe My Work

Farmer – grows crops, vegetables, and fruits

Cobbler – stitches shoes

Carpenter – works with wood to make doors, shelves, windows, etc.

Butcher – slaughters animals and sells meat

Barber – cuts hair

Construction worker – works in construction sites and performs manual labor

Find the Clue

1. Sentence from the passage: Every day they would creep into the bakery in the middle of the night and eat all the crumbs fallen down.

Move slowly and carefully

2. Sentence from the passage: They ran on the shelves, up and down, tasting all the delicious pies, cakes, and muffins.

Tasty

3. Sentence from the passage: Every day they would creep into the bakery in the middle of the night and eat all the crumbs fallen on the floor.

Small broken pieces of cake

Present Tense	Past Tense
attract	attracted
gnaw	gnawed
see	saw
give	gave
come	came
stay	stayed
walk	walked
run	ran
peep	peeped
pounce	pounced

Complete Me!

1. saw
2. haven't heard
3. wants
4. revolves

What Happened at the Pond?

Read and Answer

1. William was bored because he was alone and there was no one to play with him.
2. At the edge of the town, about five minutes from the pond
3. Because he was bored
4. He saw another boy near the pond
5. Because he had dug up her flower beds
6. It was a long walk and she didn't want him to go alone
7. They go fishing

order Me

1. It was summer holidays and William was feeling very bored
2. William asks his mother if he could go to the town to visit his friends
3. His mother forbids him from going to the town
4. Feeling very bored, William digs up the flower beds
5. William's mother scolds him for tearing up her flower beds
6. Feeling sad, William sits on the tire swing
7. He then decides to go the pond
8. William looks around silently for his mother and creeps out of the garden
9. William runs fast all the way to the pond
10. He sees a boy standing near the pond

Synonyms and Antonyms

1. synonym: till; antonym: fill

2. synonym: disappointment; antonym: happily

3. synonym: respectful; antonym: naughty

4. synonym: disobedient; antonym: obedient

5. synonym: prohibit; antonym: allow

Vegetable Patch

Story Elements

1. Cloud Mountain

2. Blacksmith

3. Pet dog

4. Radishes, turnips, carrots, beans, beets (any three)

5. Mother

Crossword Puzzle

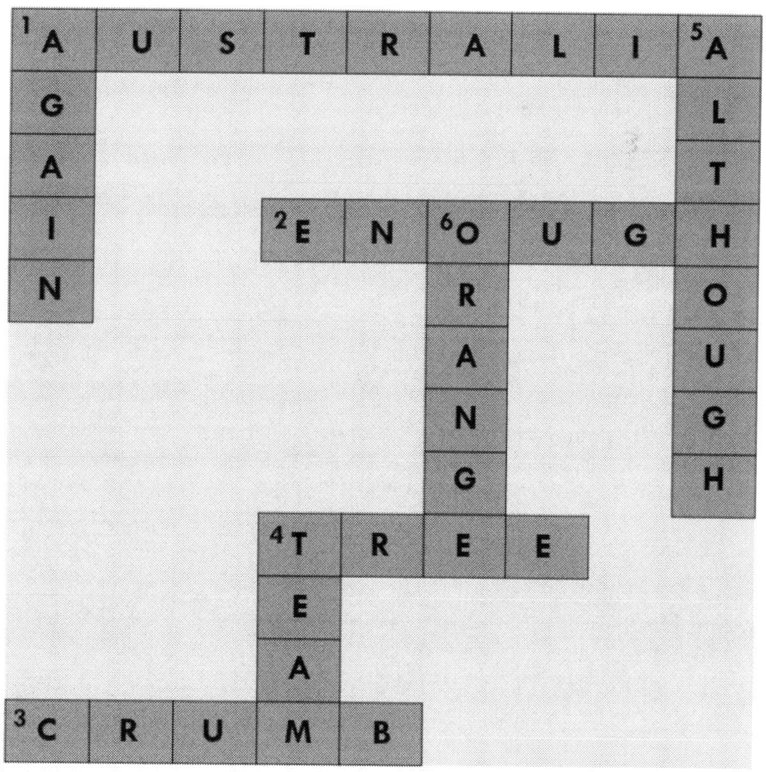

Adjectives

Green Tea; Dark Room; Fresh Vegetables; Beautiful Flower; Huge Mountain; Ugly duckling; Tall building; Dirty water; Honest Worker

Character Analysis

1. Obedient 2. Hard working 3. Impulsive 4. Helpful 5. Strong Minded

The Song of the Bee

Poem Appreciation

1. The bee
2. He goes to get honey
3. He makes wax
4. From pinks, lilies, daffodils, and columbine blossoms
5. To do nothing

Character Analysis

The right answers – hard working, happy, energetic

Make sentences – Answers may vary

Words that rhyme

yellow - fellow

clover – over

lazy – daisy

light – night

weary – dreary

The Story of How Payton Made Friends!

1. He was taller than all his classmates and he didn't like it.
2. 10 years old
3. No. He thought no one wanted to be friends with him because he was very tall.
4. Basketball
5. Answers may vary

Connecting to Text

Answer may vary

Jumbled Sentences

He would go to the last seat in the classroom and sit.

Well, here's the story of this lonely little boy called Payton.

Payton didn't like the fact that he was the tallest in his class

The teacher looked at Payton and said, "Payton, why don't you try?"

Similar Words

Start - Beginning

Truth – Fact

Choice – Select

Humiliate – Embarrass

Unscramble the Words

Enough, Simple, Worlds, Record, Oracle, Invite

A Good Play

Poem Appreciation

2. They built the ship on the staircase.

3. To play in it.

4. They built the ship with back-bedroom chairs.

5. The narrator/ R L Stevenson.

6. The narrator was alone as Tom fell out and hurt his knee.

Connecting to Text

Answers may vary

Word Connect

1. Portrait 2. Evaluate 3. Patch 4. Sample

Lion and Tiger

Compare Texts

1. Both are carnivorous and belong to the cat family.

2. Tiger is a solitary animal while lion is a social animal; Tigers are found in southeast and eastern Asia while lions are found in Africa and India.

3. Africa and India

4. Southeast and East Asia

5. The amount of pressure exerted by the jaws and teeth while biting into food (meat)

Stephen Hawking

1. 1942
2. natural sciences
3. *On the Shoulder of Giants*
4. became a professor at Cambridge
5. Both a and b

Describing Words

1. Worried; 2. tears of joy; 3. Surprised; 4. Blessed; 5. Tongue Out with eyes open; 6. Love; 7. Tongue out with eyes closed; 8. nervous; 9. Sad

Complete Me!

comfortable; healthy; fashionable; classic; genuine; positive; colorful; non-verbal

Plant Growth

Interpret Data

1. 30 days
2. Partial light
3. 8 cm
4. 2 cm
5. 32 cm, partial light

Connecting to Text

1 & 2: Answer may vary

3. water, nutrients, air, and space.

Find the Hidden objects

Coffee table, cabinet, table lamp, side table, chair, couch, picture frame, book shelf, clock, hanging lamps, carpet, plant in pot, papers, bowls, bottle

Math Answer Key

Numerical Expressions

Calculator Fun

a. 7 + 20 ÷ (36 ÷ 9) - 7

b. (16 ÷ 4) + (8 ÷ 4)

c. 27 ÷ (6-3)-3

d. 2 x (12 ÷ 3) - 1

Which Car Will Reach the Gas Station?

Car	Statement	Expression
A	Add $\frac{5}{4}$ and $\frac{7}{4}$ then multiply by 2	$(\frac{5}{4} + \frac{7}{4}) \times 2$
B	Subtract $\frac{9}{2}$ from $\frac{15}{2}$ then divide by $\frac{5}{12}$	$(\frac{15}{2} - \frac{9}{2}) \div \frac{5}{12}$
C	Sum of 8 tenths and $\frac{5}{8}$ is multiplied 8 by $\frac{2}{5}$	$\frac{2}{5} \times (0.8 + \frac{5}{8})$
D	Multiply 9 by 9 then subtract the product from 98	$98 - (9 \times 9)$
E	Divide $\frac{7}{8}$ by the difference between 12 and 5	$\frac{7}{8} \div (12 - 5)$

Car D will reach the gas station.

Simplify the Expressions

a. 67

b. 38

c. $\frac{1}{27}$

d. 2.3

e. $\frac{9}{8}$

Arrange the Markers

Marker A	$(10 \div 5) \times 2 = 4$	Marker E	$(2.5 \div 2.5) + 210 = 211$
Marker B	$(10 \div 2) + 5 = 10$	Marker F	$12 \times (102 - 5) = 1164$
Marker C	$102 - (12 \times 5) = 42$	Marker G	$(25 \times 50) + (2 \times 1) = 1252$
Marker D	$210 - (2.5 + 2.5) = 205$	Marker H	$(25 \times 52) + 1 = 1301$

The order is:

4 , 10, 42 , 205, 211, 1164, 1252, 1301

Story Problems

a. $10 - (\frac{1}{4} \times 10) - 5$

b. $5 - (\frac{1}{4} \times 5)$

What is the cost of Dinner?

Cost for 1 person:

When $n = 1$, $2 + 5n + (2n - 1) \times 2 = 9$

Cost for 5 people:

When $n = 5$, $2 + 5n + (2n - 1) \times 2 = 45$

Fill the Jars!

a. The value of $134 \times (11 + \frac{7}{8})$ is greater, so it will go into the empty jar.

b. $(2.1 + 1.45) \times \frac{3}{4}$ and $\frac{3}{4} \times (2.1 + 1.45)$, have equal values.

c. The value of $(\frac{1}{2} + \frac{1}{4}) \times 1.4$ is greater, so it will go into the empty jar.

d. The value of $95 \times (62 + \frac{7}{9})$ is greater, so it will go into the empty jar.

Blue $12 \div (4 + 2)$	The quotient of 10 and 5 times $\frac{1}{4}$	Yellow $(6 + 8) \div \frac{1}{5}$	Green $(\frac{1}{4} + \frac{3}{4}) \div \frac{7}{4}$
The sum of $\frac{1}{4}$ and $\frac{3}{4}$ divided by $\frac{7}{4}$	Triple the sum of 3.5 and $\frac{9}{8}$	Twelve divided by the sum of 4 and 2	The sum of 6 and 8 divided by $\frac{1}{5}$
Purple $\frac{1}{2} \div (\frac{5}{2} - \frac{1}{4})$	Red $(10 \div 5) \times \frac{1}{4}$	Brown $(3.5 + \frac{9}{8}) \times 3$	Half divided by the difference between $\frac{5}{2}$ and $\frac{1}{4}$

Choose the correct option

a. (10 x 29) - (6 x 29)

b. (10 x 8) - (8 x 7)

c. 120 x 10 + (100 + 5)

d. (600 x 2) +(30 - 5)

a. Expression → (180 + 135) x 30

b. Felix will deliver 9450 newspapers in 30 days.

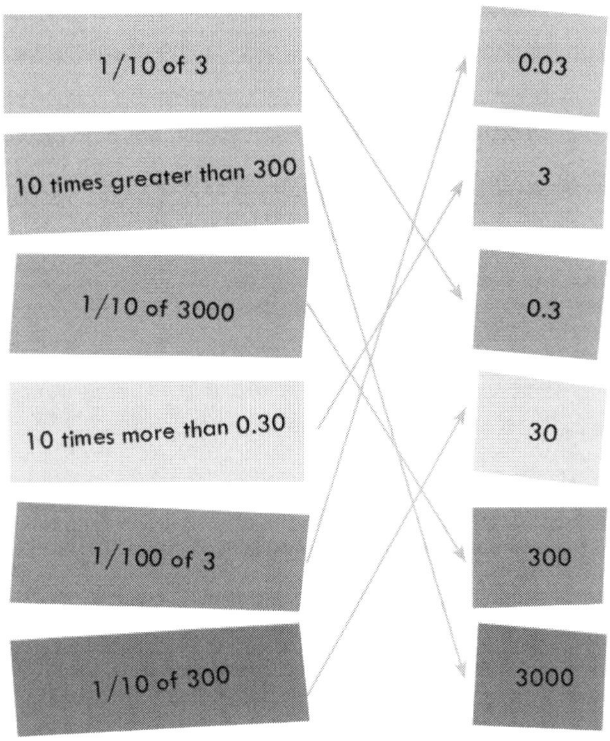

a. 1576.90

b. $7 \times 10 > 7 \times (\frac{1}{10})$

c. The value of 7 in 1576.90 is tens and that in 234.76 is tenths. Tens is greater than tenths.

Word Problems

a. (40 x 15.40) ÷ 4 = 154;
Cost of 40 pounds = $154

b. 6 x 10 = 60;
10 dollar bills measure = 60 inches

c. (200 x 156.78) ÷ 2 = 15,678;
Cost of 200 sweaters = $15,678

Compare

a. 0.6 > 0.356

b. 3 thousandths + 4 hundredths > 0.034

c. 4 x 1000 + 3 x 100 + 2 x 1/100 = 4300.02

d. 6 x 1/10 + 4 x 1/1000 < 0.640

e. 2 tens 2 tenths > 2.02

a. 345.678

Expanded form = 3 x 100 + 4 x 10 + 5 x 1 + 6 x (1/10) + 7 x (1/100) + 8 x (1/1000)

Word form = Three hundred forty five six hundred seventy eight thousandths

b. 112.345

Expanded form = 1 x 100 + 1 x 10 + 2 x 1 + 3 x (1/10) + 4 x (1/100) + 5 x (1/1000)

Word form = One hundred twelve three hundred and forty five thousandths

c. 124.009

Expanded form = 1 x 100 + 2 x 10 + 4 x 1 + 9 x (1/1000)

Word form = One hundred twenty four nine thousandths

a. 16.87 x 100 = 1687

thousands	hundreds	tens	ones	.	tenths	hundredths
		1	6	.	8	7
1	6	8	7	.	0	0

b. 6.87 x 10 = 68.7

thousands	hundreds	tens	ones	.	tenths	hundredths
			6	.	8	7
		6	8	.	7	0

c. 687 ÷ 10 = 68.7

thousands	hundreds	tens	ones	.	tenths	hundredths
	6	8	7			
		6	8	.	7	

a. Actual area of the neighborhood = 0.451 x 1000 = 451 square meters

thousands	hundreds	tens	ones	.	tenths	hundredths	thousandths
				.	4	5	1
	4	5	1	.	0	0	0

b. Actual area of the park = 451/10 = 45.1 square meters

Exponential and Standard Form

a.

Standard Form	Exponential Form
10,000	10^4
1,000	10^3
10 x 1000	10^4
1,000 x 1,000	10^6
100,000	10^5

b.

Exponential Form	Standard Form
10^3	1,000
$8700 \div 10^4$	0.87
3.808×10^2	380.8
$76 \div 10^4$	0.0076
9×10^5	900,000

Number Patterns

a. **0.63**, 6.3, 63, 630, **6300, 63000**

b. **98760**, 9876, 987.6, **98.76**, 9.876, **0.9876**

c. 7,560,000, 75,600, **756**, 7.56, **0.0756**

d. 0.009, 0.9, 90, **9000, 900,000, 90,000,000**

Help Defective Jones

a. 1000 millions

b. 1000 times thousands

c. 0

d. 1,110,110

e. 989,989

a. 1,500,000

b. 90,000

c. 4,048

d. 15,600

e. 469,335

f. 52,584

Words	Equation	Solution
20 times the sum of 12 and 3	**20 x (12 + 3)**	**300**
Sum of 1400 and 1300 divided by 100	(1400 + 1300) ÷ 100	**27**
14 times the difference between 567 and 230	14 x (567 - 230)	**4,718**
Divide the difference between 1236 and 1200 by 3	**(1236 - 1200) ÷ 3**	**12**
23 times the sum of 149 and 120	**23 x (149 + 120)**	**6,187**

Aaron's Having a Party

42 packs of chips = 42 x 3.45 = $144.90

36 cans of juice = 36 x 1.25 = $45

Total = 144.90 + 45 = $189.90

Ryan's Math Assignment

a. 48 x 50 = 2400

b. 4.8 x 50 = 240

Similarity: The answers have the same digits

Difference: The place value of the digits are different

Reasoning: The answer in (b) is smaller than the answer in (a) because ones are ten times more than tenths. So (a) is ten times bigger than (b)

The Missing Divisor

a. Divisor - 10 ; exponent - 10^1

b. Divisor - 100 ; exponent - 10^2

c. Divisor - 1000 ; exponent -10^3

d. Divisor - 10 ; exponent - 10^1

Equation	 Or 	Reasons
a. 5 tens = 3 tens + 2 tens		**5 tens = 5 x 10 = 50 3 tens + 2 tens = 3 x 10 + 2 x 10 = 30 + 20 = 50 So, they are equal**
b. 53 x 100 x 8 = 530 x 8 x 10		**53 x 100 x 8 = 5300 x 8 = 42, 400 530 x 8 x 10 = 4,240 x 10 = 42,400 So, they are equal**
c. 67 x 90 x 100 = 67 x 9 x 10 x 10 x 10		**67 x 90 x 100 = 603,000 67 x 9 x 10 x 10 x 10 = 603,000 So, they are equal**
d. 50 tens x 6 hundreds = 60 tens x 5 hundreds		**50 tens x 6 hundreds = 500 x 600 = 300,000 60 tens x 5 hundreds = 600 x 500 = 300,000 So, they are equal**
e. 5 x 78 x 3000 = 50 x 780 x 3		**5 x 78 x 3000 = 1,170,000 50 x 780 x 3 = 117,000 So, they are not equal**

a. 6,453

b. Divided, 100

c. 300 inches

d. 98.5

How Much Will Angela Save?

a. Angela will have $19,440 by 3rd year

b. It will take her approximately 6 years to save $36,000

a. 7 cups of cocoa powder

b. 359.71 grams

c. The first batch weighs more than the second by 62.69 grams

d. 96.19 grams

e. 1,690 ounces

b. $0.8 \times 0.1 = 0.08$

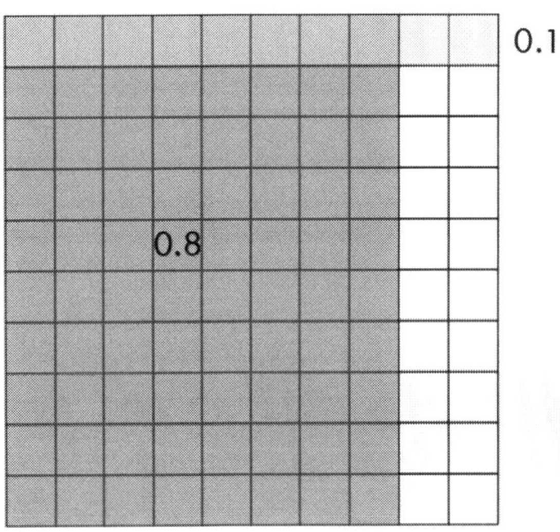

c. $0.4 \times 0.6 = 0.24$

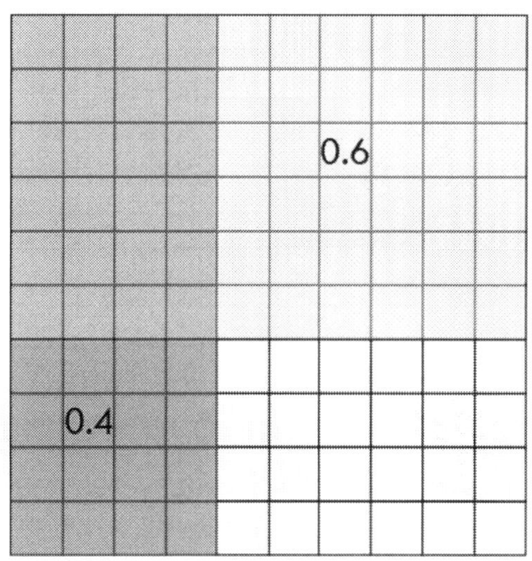

d. $1.3 \times 0.2 = 0.26$

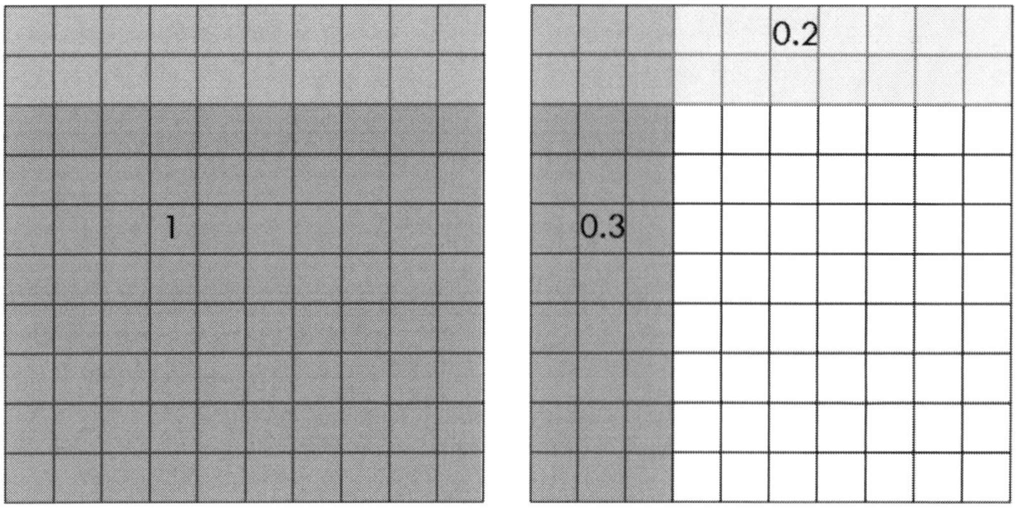

Number operations in Fractions

Sum of the whole number and fraction is mixed fraction.

a. $1\frac{2}{7}$ b. $4\frac{4}{7}$ c. $3\frac{1}{2}$ d. $2\frac{6}{9}$

Equivalent Fractions

a. $\frac{3}{7} + \frac{3}{7}$

b. $\frac{5}{4} + \frac{5}{4} + \frac{5}{4}$

Fraction on Number Line

a.

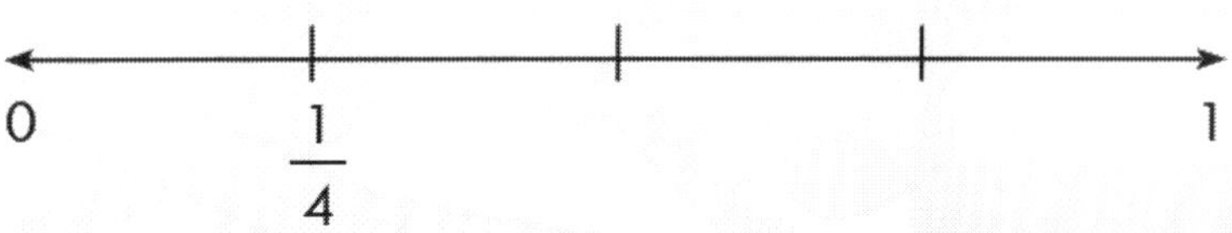

$\frac{1}{4} + \frac{1}{4} + \frac{1}{4} + \frac{1}{4} = \frac{4}{4} = 1$

b.

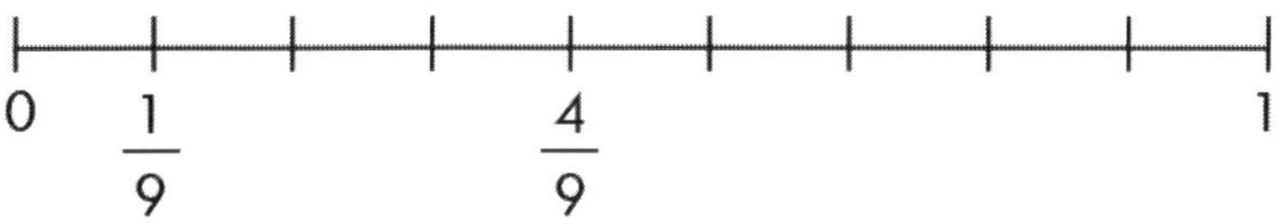

$$\frac{1}{9} + \frac{4}{9} = \frac{5}{4}$$

c.

$$\frac{2}{7} - \frac{2}{7} = 0$$

d.

$$\frac{3}{5} - \frac{1}{5} = \frac{2}{5}$$

Fraction Model

a. $\frac{7}{6}$ b. $3\ \frac{1}{5}$ c. $1\ \frac{19}{20}$ d. $\frac{1}{12}$ e. $\frac{5}{6}$

Greater Than or Less Than 1

a. Less than one

b. Greater than one

c. Greater than one

a.

$$\frac{1}{4} + \frac{7}{4} + 2\ \frac{2}{3} + \frac{1}{3}$$

$$= \frac{8}{4} + 2\ \frac{3}{3}$$

$$= \frac{8}{4} + 3$$

$$= 3\ \frac{8}{4}$$

$$= 5$$

b.

$$\frac{5}{6} - \frac{1}{6}$$

$$= \frac{4}{6}$$

c.

$$4\ \frac{3}{7} - \frac{1}{7}$$

$$= 4$$

d.

$$2\ \frac{3}{5} + \frac{2}{5}$$

$$= 3$$

Make the Statements True

a. 8 b. $\frac{3}{12}$ c. $30\frac{1}{4}$

a. The total length of 5 boards if $4\frac{1}{2}$ m.

b. Nancy took $\frac{13}{36}$ hours more than Tom.

c. Carl stretched for $5\frac{2}{3}$ minutes.

d. Samuel needs $3\frac{9}{40}$ meters of fabric.

Justify Eric's Solution

$3 - 1\frac{1}{3} = 1\frac{2}{3}$, thus Eric is wrong. Drawing will vary.

Is Elise Right?

$5\frac{3}{4} + 2\frac{1}{4} = 8$, thus her answer is wrong. Drawing will vary.

Measuring Length, Mass, and Capacity

Mr. Rod Has only a Centimeter Tape.

Convert the measurement given in meters and millimeters to centimeters.

$\frac{9}{10}$ m = 90 cm 550 mm = 55 cm

Convert to Like Units

a. 1.25 liters or 1250 ml of juice

b. 2.75 minutes

c. Olivia's rope piece is longer than Kristy by 9 cm.

Compare the Measurements

a. = b. > c. <

Water Consumption of Miley and Lilly

a. Amount of water used by Miley - 74 gallons

Amount of water used by Lilly - 294 quarts = 73.5 gallons

b. Lilly uses more water to wash her car.

c. Yes, Miley will be using less water than Lilly. $\frac{2}{3}$ of 2 gallons is 1.333 gallons. Miley is using 2 - 1.333=0.667 gallons of water, while Lilly is using 1 gallon of water.

How Much Ribbon?

a. Lisa needs 112.5 feet of ribbon.

b. Lisa would have to pay $11.25

c. 25 x 100 x 1.5 yards or 25 x 100 x 1.5 x 3 feet

Volume of Solid Shapes

There are six 1 cm cubes, so the volume is 6 cubic cm and not 5 cubic cm, thus Demi is wrong.

Find the Volume

a. 36 cubic units

b. 20 cubic cm

c. 27 cubic cm

Predict the Number of Cubes

a. 5 cubes

b. 2 cubes

Drawings may vary.

Guess the Number of Cubes

8 unit cubes will fit inside the box.

Robert and Jamie's Argument

Robert is right. 16 cubes to be added 6 times or 24 cubes added 4 times will be the volume of the prism.

Volume of one layer is 25 cubic inches Volume of two layers is 50 cubic inches Volume of five layers is 125 cubic inches

a. 24 cubic in
b. 36 cubic cm
c. 1152 cubic m
d. 128 cubic ft

a. The volume of the cupboard is 24,192 cubic inches.
b. The volume of the box is 224 cubic centimeters.
c. The height of the container will be 2 cm.
d. He can build a box using the dimensions - 12 inches x 3 inches x 2 inches

a. 74 cubic cm
b. 392 cubic cm
c. 420 cubic ft

Length: 2 mm
Width: 2 mm
Height: 4 mm
Volume: 16 mm^3
2 mm x 2 mm x 4 mm = 16 mm^3

Line Plots

Fund-raising Rally Plot

a. 3 students
b. 4 students
c. 20 students
d. 4 miles
e. 1 student

a. 26 pieces
b. Minimum length - 1 inch and maximum length - 3 inch
c. 2 inches
d. She will have two 1-inch pieces
e. 3 inches + 1 $\frac{1}{4}$ inch + 3 inches + 4 $\frac{12}{4}$ inches + 4 inches + 6 $\frac{3}{4}$ inches + 12 $\frac{1}{2}$ inches + 4 $\frac{6}{4}$ inches + 12 inches
= 55 inches

Laura's Reading Schedule

a. 2 hours

b. 1 hour

c. 8 days

d. 25 days

e. $33\frac{3}{4}$ hours

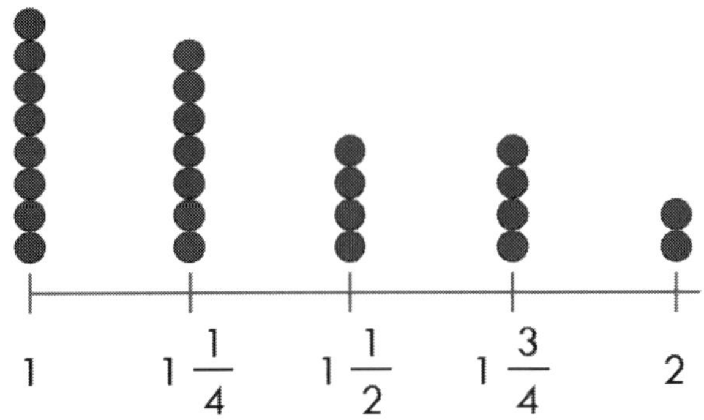

Did Sean Collect Enough Water?

Amount of water collected = $\frac{1}{8} + \frac{2}{8} + \frac{4}{8} + \frac{4}{8} + \frac{4}{8} + \frac{6}{8} + \frac{6}{8} + \frac{7}{8} + \frac{8}{8} = \frac{42}{8} = 5\frac{2}{8}$. Yes he was able to collect enough water to reuse.

Fair Share of Pudding?

Total amount of pudding = $\frac{1}{8} + \frac{1}{8} + \frac{2}{8} + \frac{3}{8} + \frac{3}{8} + \frac{4}{8} + \frac{6}{8} + \frac{8}{8} = \frac{28}{8} = 3\frac{1}{2}$ cups of pudding.

Redistribution of pudding = $\frac{7}{2} \div 8 = \frac{7}{16}$ cups of pudding.

Thus, after the pudding is redistributed fairly, Leo and his friends will each have $\frac{7}{16}$ of a cup of pudding.

Geometry

Answers may vary. Sample trapezoid with no right angles and 2 obtuse angle is given below

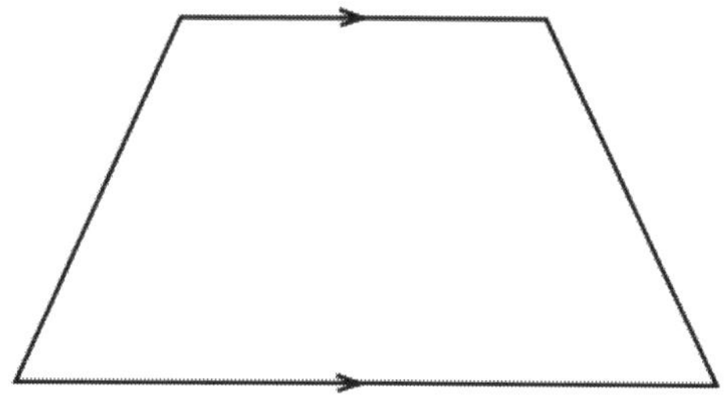

The drawing

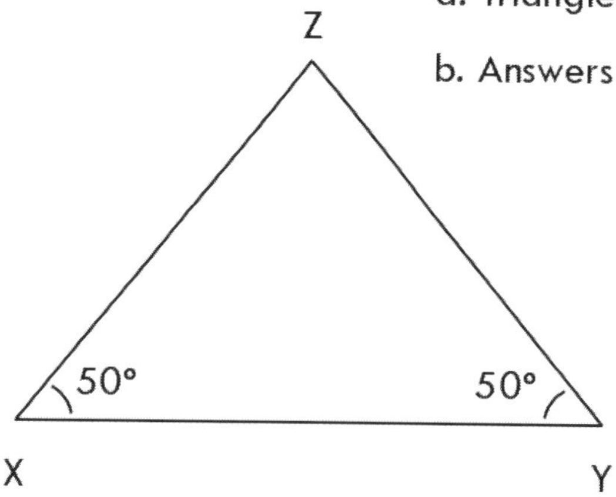

a. Triangle

b. Answers may vary.

217 ÷ 4 = 54.25 cm.

Therefore, each side will measure 54.25 cm

Answers may vary.

a. Properties of a kite:

- It is a quadrilateral
- It has two pairs of adjacent sides that are equal.
- The diagonals are perpendicular to each other.

b. When both pairs of sides are equal or when it is a rhombus.

No, the shape is not a quadrilateral. Explanations may vary

Properties of a rhombus

- It is a quadrilateral
- All four sides are of equal length
- Opposite sides are parallel
- The diagonals are perpendicular bisectors
- The angles next to each other add up to 180°

Can You Convert a Rectangle Into a Square?

a & b Students may change the rectangle to sides of 8 cm each or 5 cm each.

Geometry Quiz

a. An isosceles trapezoid is one which has one pair of parallel sides and another pair of equal sides.

b. A trapezoid that has only one pair of parallel sides.

c. When all four sides of the rectangle are equal.

d. Square

a. All **rectangles** are **parallelograms** but all parallelograms are not **rectangles**.

b. All **rhombuses** are **kites** but all **kites** are not **rhombuses**.

Drawing will vary

Attributes:

- It is a quadrilateral with opposite sides parallel to each other
- The opposite sides are equal in length
- It has two pairs of angles that add to 180°
- Their opposite angles are equal in measure
- The diagonals bisect each other into two equal halves

Function Table

X	0	2	4	6
Y	0	4	8	12

X	0	3	6	9
Y	0	6	12	18

X	0	5	10	15
Y	0	1	2	3

X	0	3	6	9
Y	0	2	4	6

A(2,2)

B(4,5)

C(5,3)

D(6,2)

E(7,7)

A Fruity Grid

Pomegranate	F, 3
Apple	B, 6
Mango	A, 3
Lemon	C, 5
Pineapple	D, 2
Watermelon	E, 4
Guava	G, 4
Orange	E, 7

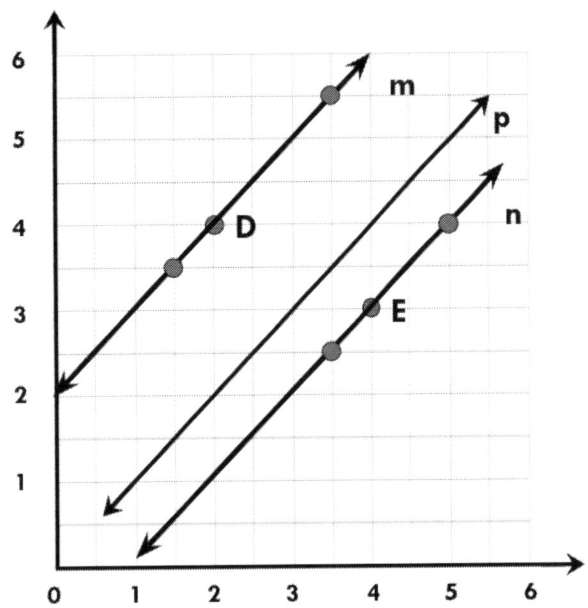

b. $(3\frac{1}{2}, 2\frac{1}{2})$; (5, 4)

c. Rule: y is one less than x

e. $(1\frac{1}{2}, 3\frac{1}{2})$; $(3\frac{1}{2}, 5\frac{1}{2})$

f. Rule: y is 2 more than x.

g. Line m has y- coordinates that are 3 more than line n's coordinates. Both the lines m and n are parallel to line p.

Line m
Rule: y is three times x

x	y	(x,y)
0	0	(0,0)
1	3	(1,3)
2	6	(2,6)
3	9	(3,9)

Line n
Rule: y is three times x, and then add 1

x	y	(x,y)
0	1	(0,1)
1	4	(1,4)
2	7	(2,7)
3	10	(3,10)

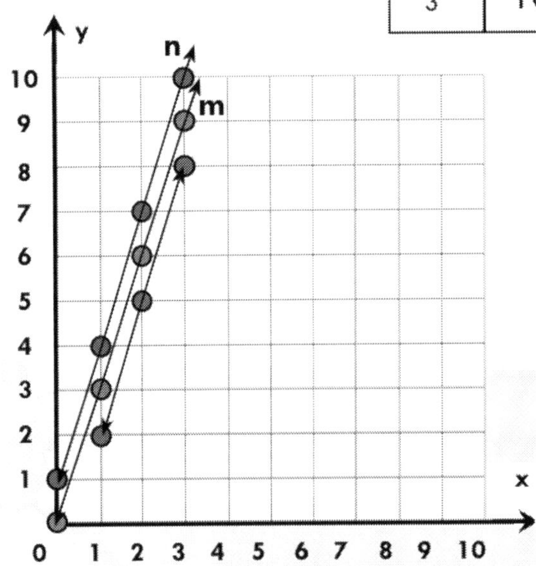

b. Line n has y-coordinates that are 1 more than line m.

c. Line m will have y-coordinates that are 1 less than line n.

a.

x	0	1	2	3	4	5
y	8	9	10	11	12	13

$y = x +$ 8

b.

x	1	3	8	5	11	12
y	5	7	12	9	15	16

$y = x +$ 4

c.

x	2	3	9	5	6	8
y	8	12	36	20	24	32

$y =$ 4X

d.

x	1	5	7	3	13	6
y	0	4	6	2	12	5

$y =$ X – 1

e.

x	4	6	8	12	24	16
y	2	3	4	6	12	8

$y =$ X/2

Tour Around the City

a. Store

b. Library

c. Art gallery

d. School

a. C(5,9)

b. A(5,3) B(2,6) C(5,9)

c. D(8,6)

d. Square

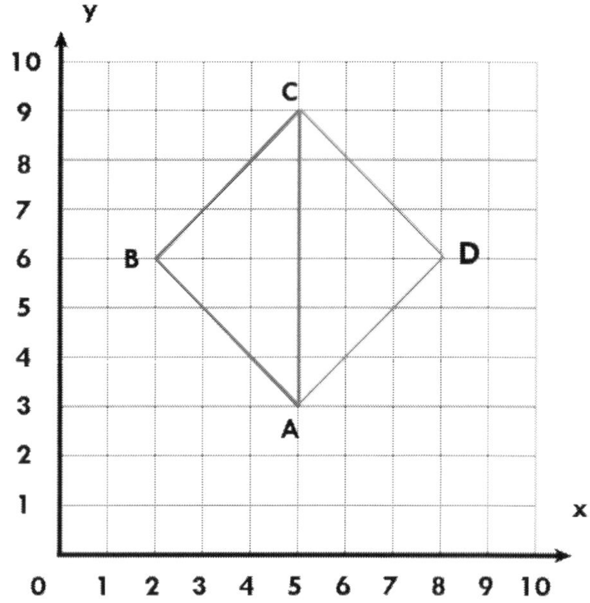

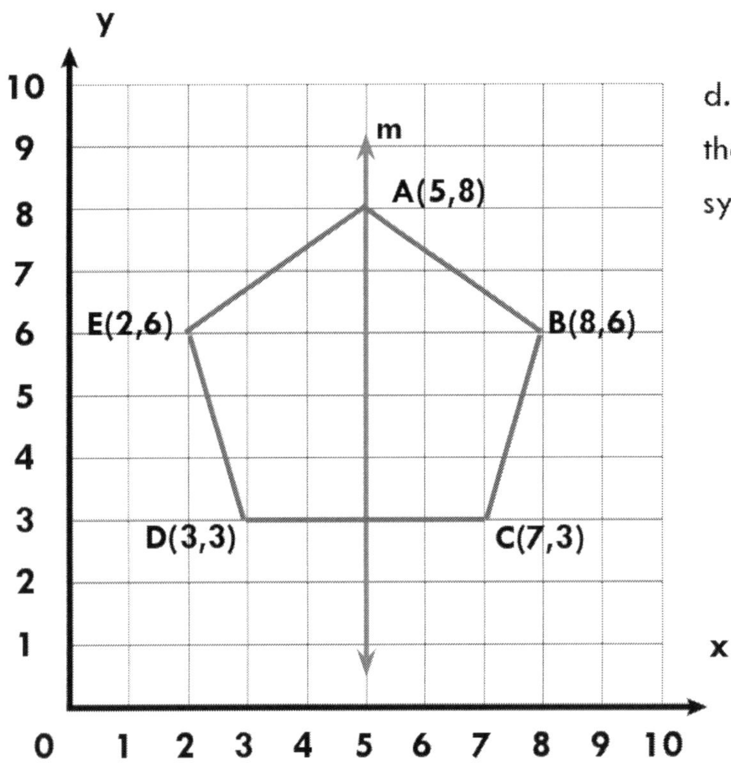

d. Based on the coordinates of the pentagon the rule for line of symmetry will vary.

Richard's Daily Travel Plan

Saturday - 9.5

Rule - 2 + x(1.5)

Types of Matter

objects and Elements

- Foil - aluminium
- Pencil lead - carbon
- Diamond - carbon
- Wedding rings - gold
- Wire - copper
- Filament - tungsten

Types of Atoms in a compound

- Salt - sodium and chlorine
- Water - hydrogen and oxygen
- Sugar - carbon, hydrogen, and oxygen
- Carbon dioxide - carbon and oxygen
- Mouthwash - hydrogen and oxygen
- Chalk - calcium, carbon and oxygen

Metals Vs Nonmetals

	Physical Properties	Definition	Metals	Nonmetals
1.	Malleability	Ability to be hammered, beaten or pressed into thin sheets	Highly malleable Except Zinc, Arsenic and Antimony.	Non malleable Nonmetals are brittle and can be broken easily.
2.	Sonority	Property of producing a ringing sound when struck with a hard substance.	Sonorous	Non sonorous
3.	Ductility	Ability to form thin wires	Ductile Example Copper, Aluminium and Iron	Non ductile because the nonmetals are brittle.

4.	Lustre	Substance having a shiny surface.	Lustrous, like Gold, Copper, and Silver	Non lustrous. Nonmetals are dull and do not have lustre.
5.	Hardness	Property of making cutting substances difficult.	Metals are hard.	Nonmetals are brittle and can be broken easily. Hence, not hard.
6.	Thermal and electrical conductivity	The ease with which electric charge or heat can pass through a substance.	Good conductors of heat and electricity.	Bad conductors of heat and electricity.
7.	Density	A measurement that compares the amount of matter a substance has to its volume.	High density	Low density
8.	Melting and boiling points	Temperature at which the substance changes from solid to liquid and vice versa.	Usually melt/ boil at higher temperatures than nonmetals.	melt/boil at lower temperatures.

Methods of Separation

- Magnetic separation
- Evaporation
- Handpicking
- Sieve
- Handpicking
- Sedimentation
- Filtration
- Handpicking
- Evaporation

Crossword

1. Dissolves
2. Sieve
3. Diffusion
4. Distillation
5. Solvent
6. Residue
7. Filtrate
8. Sublimation
9. Sedimentation
10. Evaporation

Properties of Matter

Physical and Chemical Properties

1. Physical reaction
2. Physical reaction
3. Chemical reaction
4. Chemical reaction
5. Physical reaction
6. Physical reaction
7. Chemical reaction
8. Chemical reaction
9. Chemical reaction
10. Chemical reaction
11. Chemical reaction
12. Physical reaction
13. Physical reaction
14. Chemical reaction
15. Chemical reaction

Physical Vs Chemical Changes

1. Physical change
2. Physical change
3. Chemical change
4. Physical change
5. Physical change
6. Physical change
7. Chemical change
8. Physical change
9. Chemical change
10. Physical change
11. Chemical change
12. Physical change
13. Chemical change
14. Chemical change
15. Chemical change

Elements and Symbols

1. a. C - Carbon

b. H - Hydrogen

c. Al - Aluminium

d. B - Boron

e. I - Iodine

f. Au - Gold

g. Ag - Silver

h. Fe - Iron

i. Pb - Lead

j. He - Helium

2. a. Argon - Ar
b. Oxygen - O
c. Zinc - Zn
d. Uranium - U
e. Platinum - Pt
f. Silicon - Si
g. Lithium - Li
h. Sulfur - S
i. Fluorine - F
j. Sodium - Na

Periodic Table crossword

Across:

4. Element
7. Cesium
8. 118
9. Atomic number
13. Proton
14. Hydrogen
16. Period

Down:

1. Oxygen
2. Seven
3. Atom
5. Electron
6. 18
7. Compound
10. Mendeleev
11. Group
12. Molecule
14. Helium
15. Neutron

Around the World on a Periodic Table

a. Pakistan
b. Peru
c. China
d. Iceland
e. Netherlands
f. Spain
g. Singapore
h. Britain
i. France
j. United States of America
k. Poland
l. Argentina

Element Puns

1. Iron
2. Tungsten
3. Argon
4. Antimony
5. Iodine
6. Mercury
7. Uranium
8. Neon
9. Copper
10. Californium
11. Sodium
12. Samarium

Burning of Matter

A. Pure element

B. Mixture of compound and element

C. Mixture of two elements

D. Pure compound

E. Mixture of two compounds

No.	Column A	Column B	Column C
1	Lemonade	Mixture	Boiling point, evaporation
2	Italian Dressing	Mixture	Density, solubility
3	Salt and Pepper	Mixture	Density, solubility
4	Iron filings and sand	Mixture	Magnetism, density
5	Silver	Element	
6	Pepsi	Mixture	Boiling point, evaporation
7	Steam	Compound	
8	A burrito	Mixture	Size
9	Toothpaste	Compound	
10	Chicken Soup	Mixture	Boiling point, evaporation, size

Types of Substances

1.

Image No.	Description
B	Mixture of a compound and an element
C	Mixture of two elements
D	Pure compound
E	Mixture of two compounds
A	Pure element

2.

No.	Column A	Column B	Column C
1	Lemonade	Mixture	Boiling points
2	Italian Dressing	Compound	
3	Salt and Pepper	Mixture	Solubility
4	Iron filings and sand	Mixture	Magnetic property
5	Silver	Element	
6	Pepsi	Compound	
7	Steam	Compound	
8	A burrito	Mixture	Size of matter
9	Toothpaste	Compound	
10	Chicken Soup	Mixture	Size and boiling point

Life Sciences

Plants vs Animals

Characteristics	Plants	Animals
Eat		✓
Reproduce	✓	✓
Grow	✓	✓
Make food	✓	
Need air	✓	✓
Respond to stimuli	✓	✓
Form seeds	✓	
Have chlorophyll	✓	
Need sunlight	✓	✓
Move on land		✓
Can fly		✓

Plant cell and Animal cell

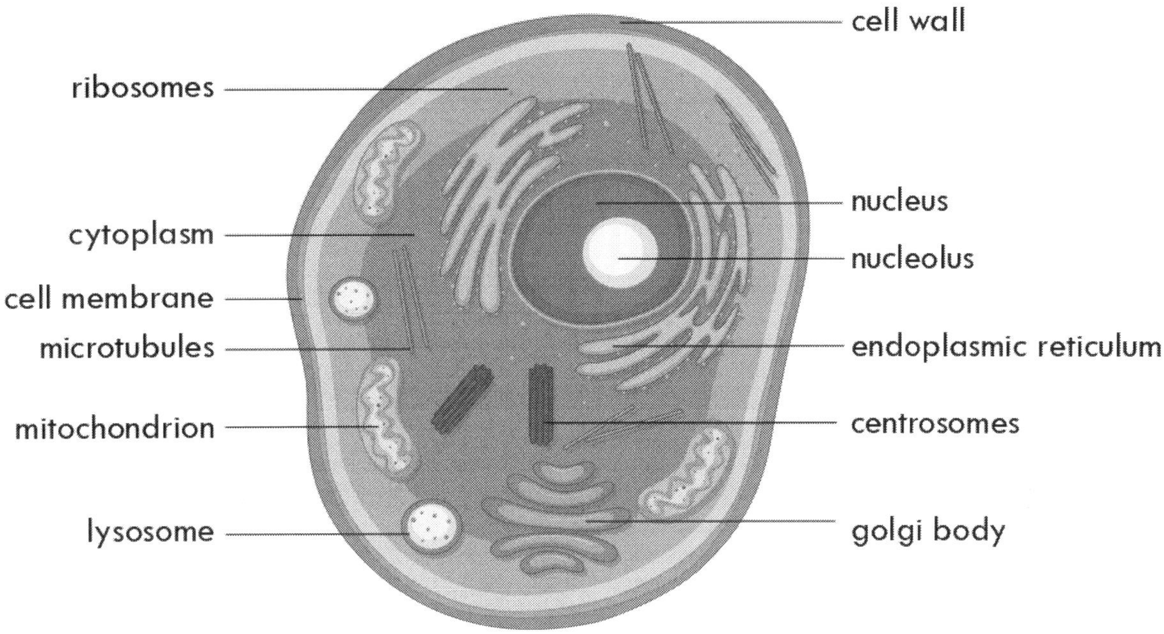

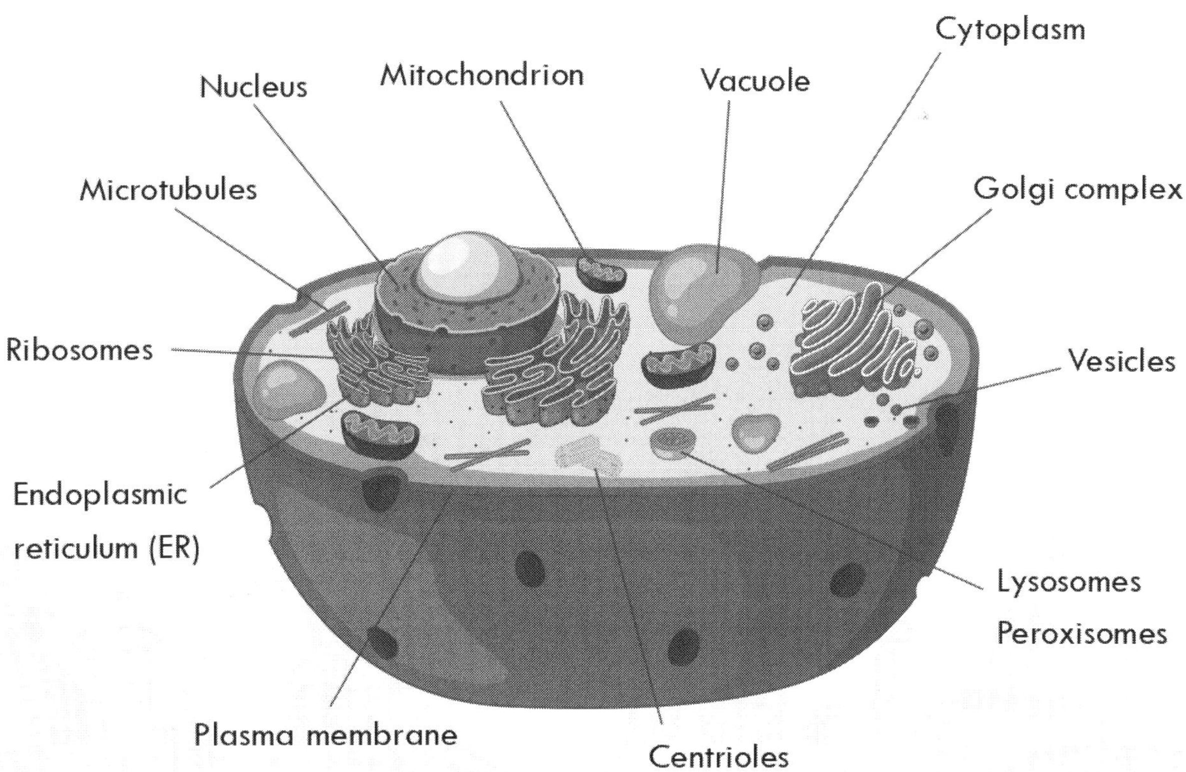

Systems in our Body

1. Digestive system
2. Skeletal system
3. Circulatory system
4. Respiratory system

Human Heart

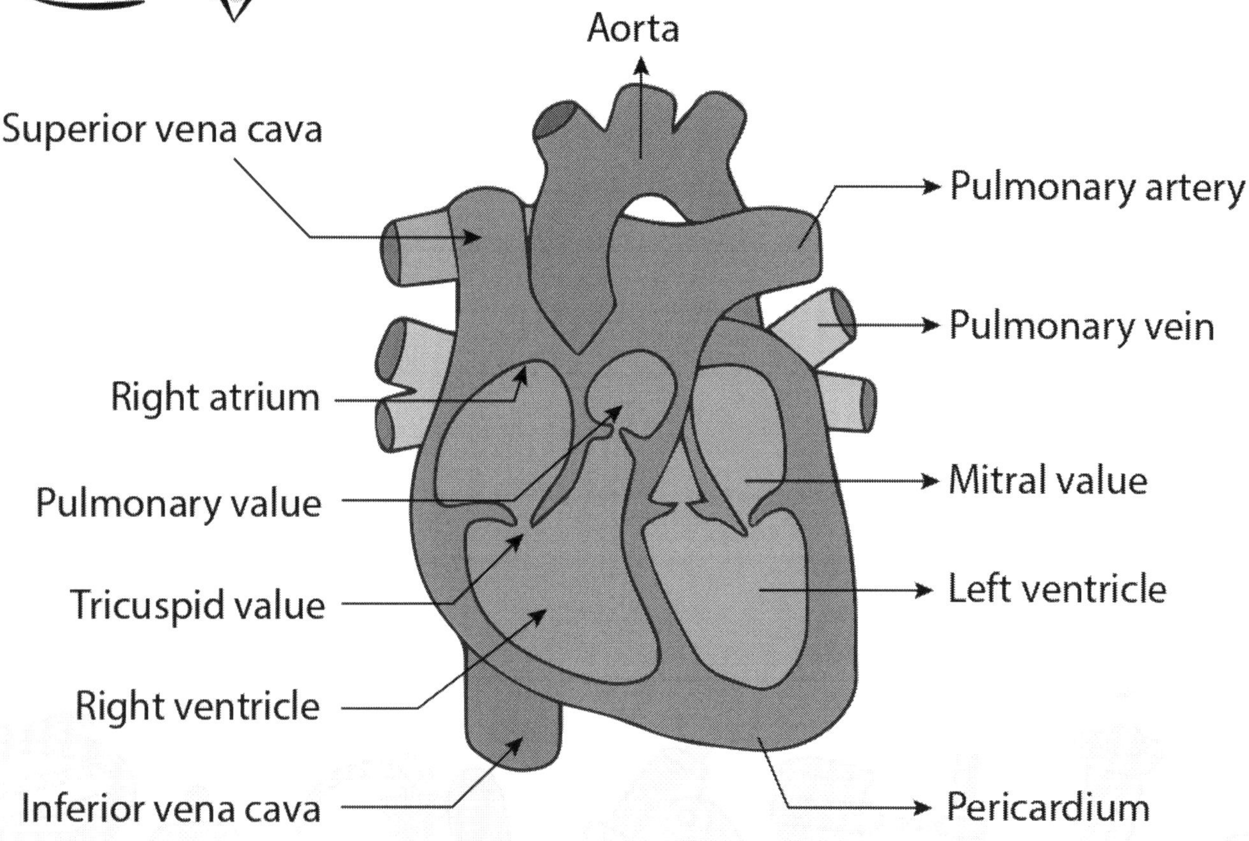

Circulatory System

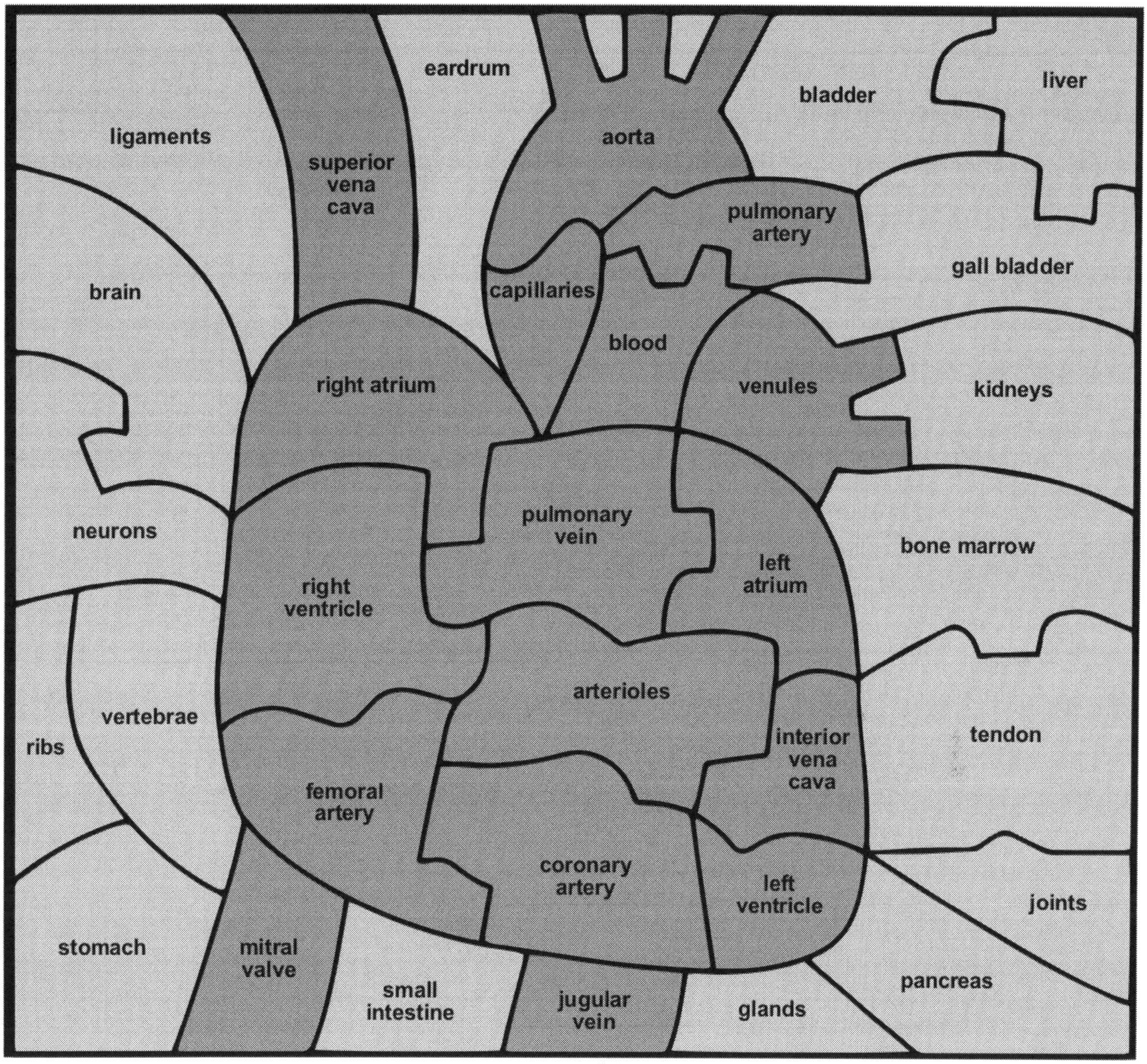

Steps in Digestion

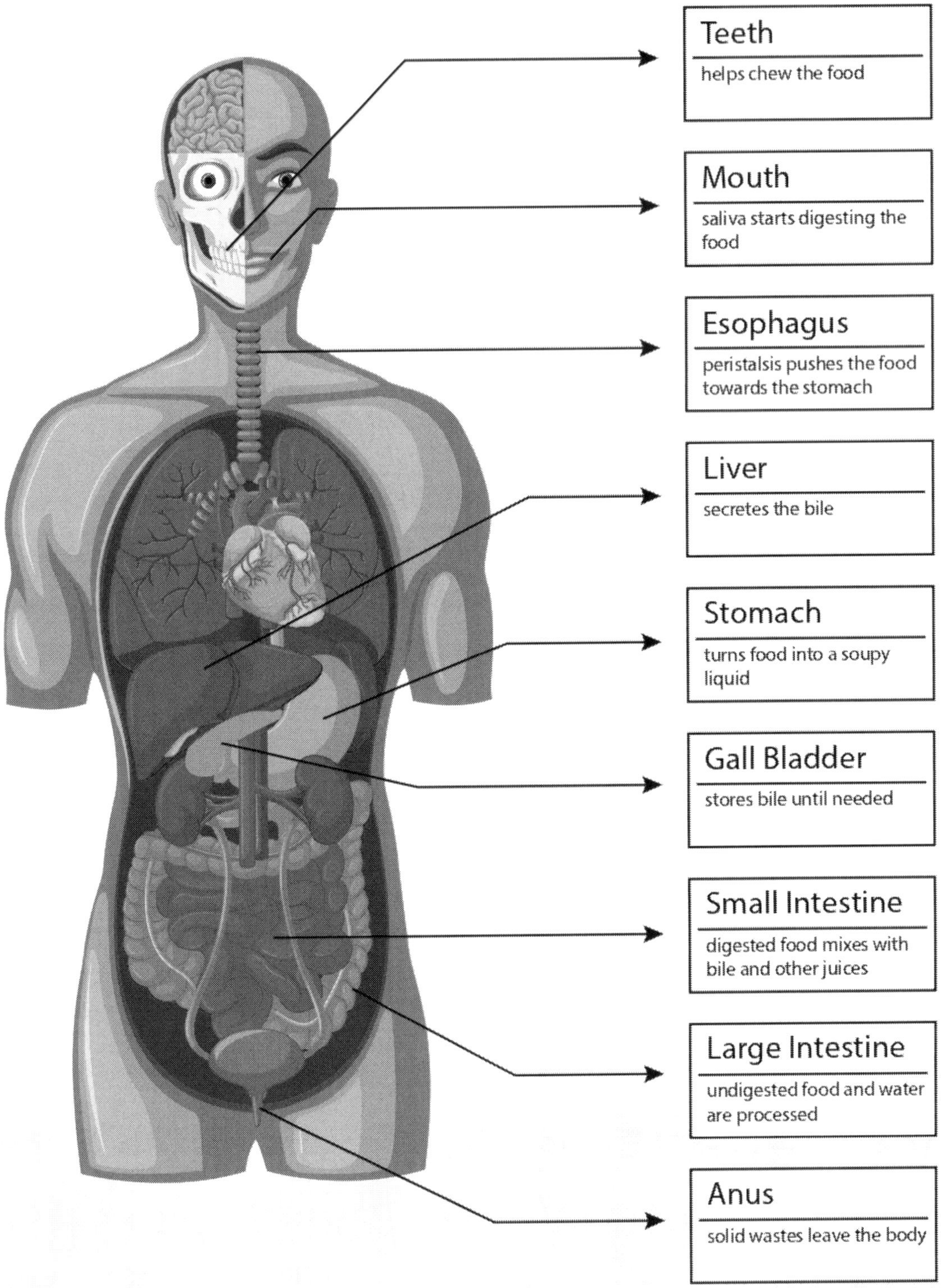

Urinary System

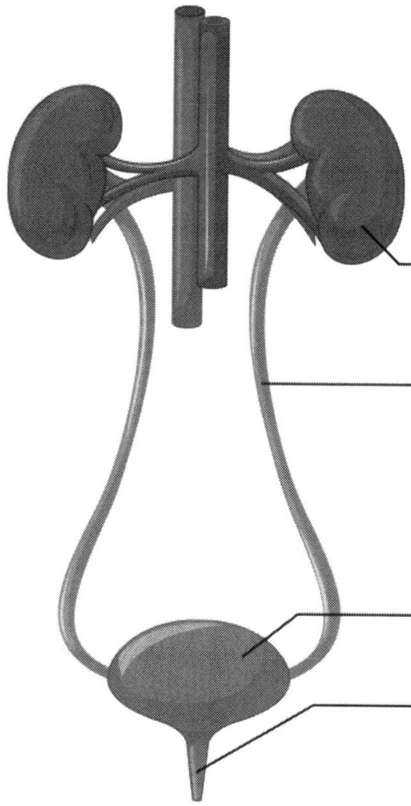

Functions:

1. Kidneys -

a. regulating and filtering minerals from blood

b. filtering waste materials from food ingested

2. Ureters -

a. carry urine from the kidneys into the bladder

3. Bladders -

a. stores urine to allow infrequent urination

4. Urethra -

a. carries urine from the bladder to outside the body

Human Kidney

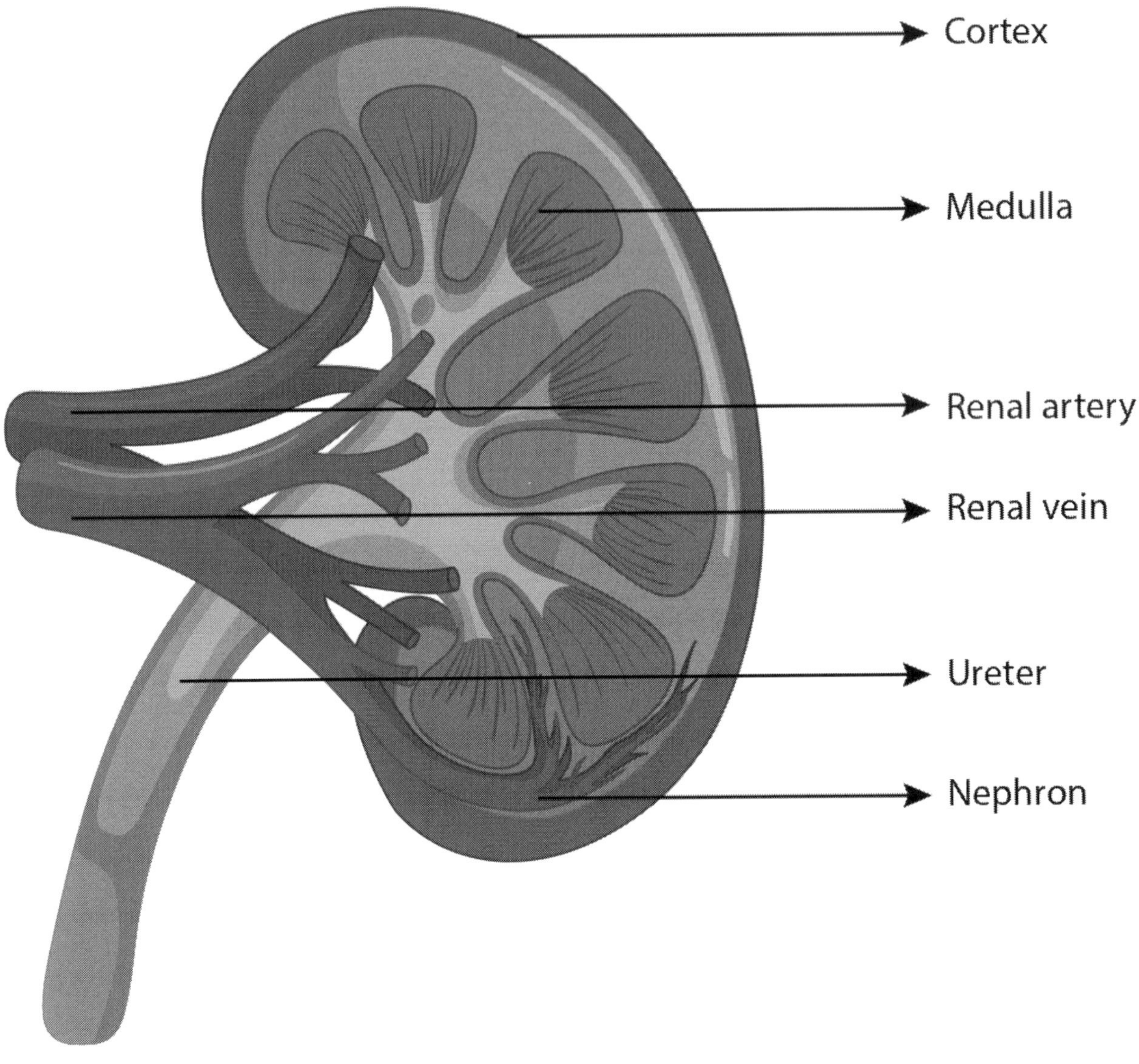

Respiratory System

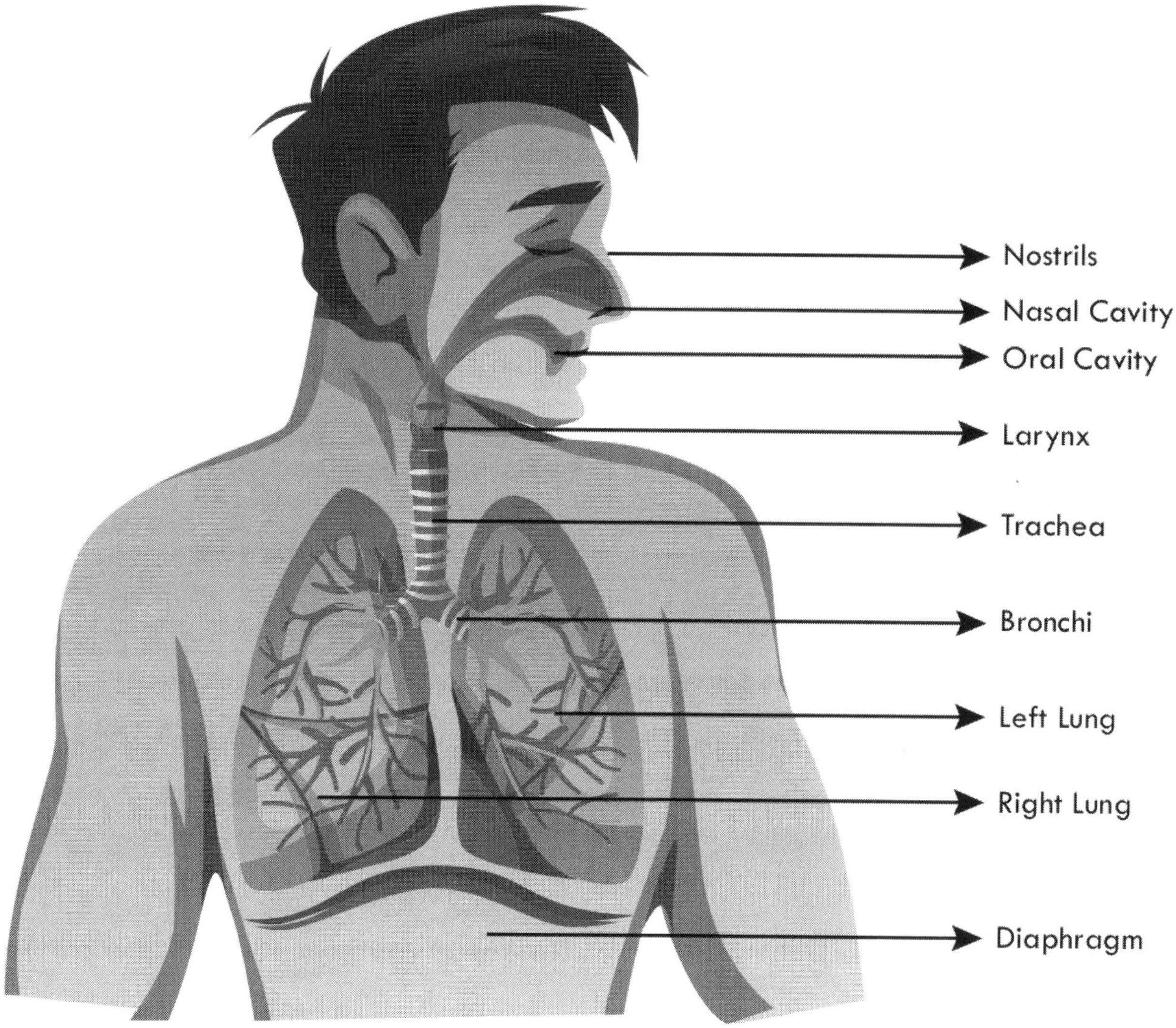

3.

- alveoli
- cartilage
- inhale, exhale
- voice box
- diaphragm
- pleural fluid
- trachea

Exchange of Gases

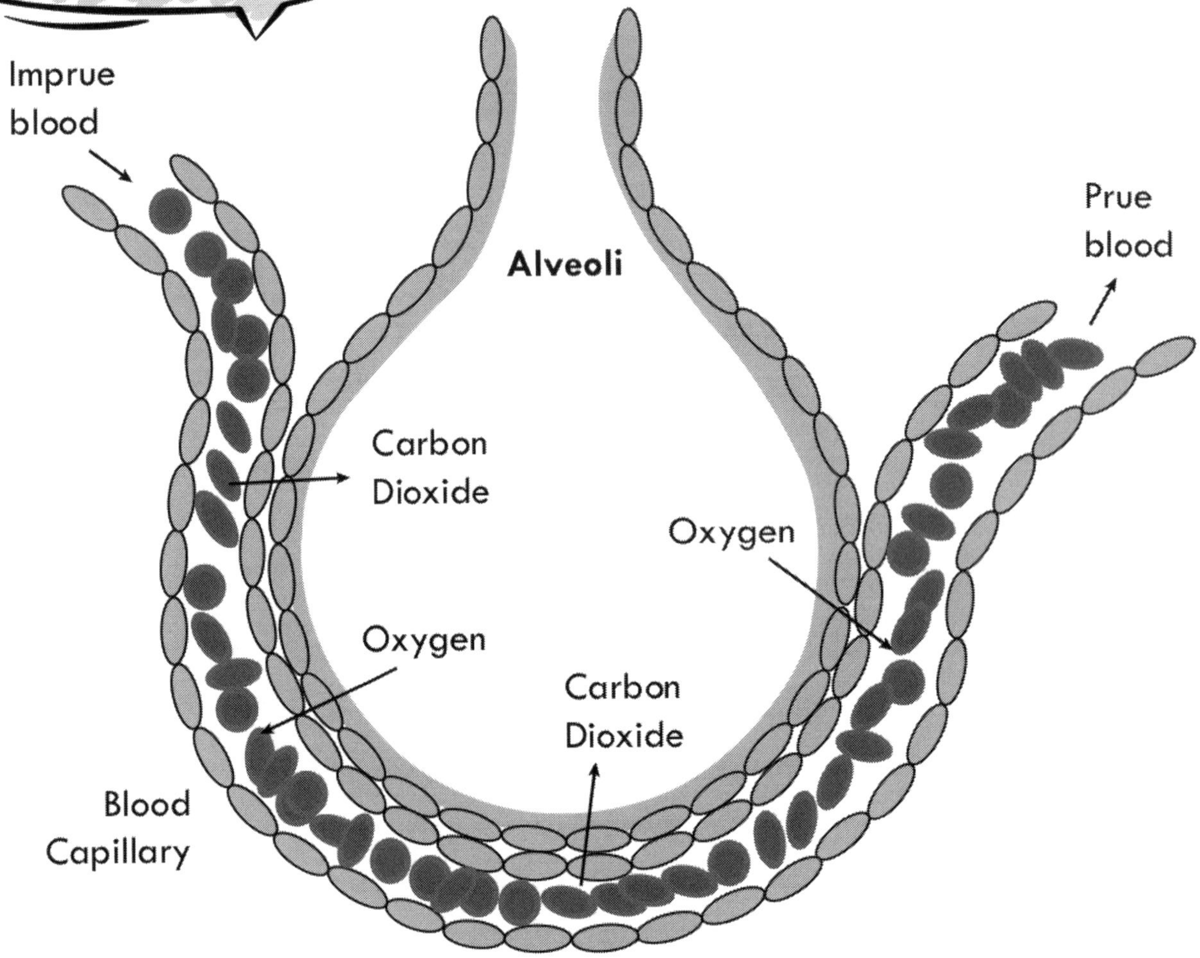

Photosynthesis and Cellular Respiration

1. $6CO_2 + 6H_2O \xrightarrow{\text{light}} C_6H_{12}O_6 + 6O_2$

Carbon dioxide | Water | Sugar | Oxigen

2. **Cell Respiration Formula**

$$C_6H_{12}O_6 + 6O_2 \longrightarrow 6CO_2 + 6H_2O + ATP$$

Glucose | Oxygen | Carbon dioxide | Water | Energy

Earth Sciences

Pop Quiz

1. Uranus
2. Saturn
3. Neptune
4. Venus
5. Earth
6. Saturn
7. Mars
8. Venus
9. Jupiter
10. Mercury
11. Jupiter
12. Venus
13. Mars
14. Sun
15. Saturn, because it has 82, 3 more than Jupiter's 79.

Fact or opinion?

1. Opinion
2. Fact
3. Opinion
4. Opinion
5. Fact
6. Fact
7. Opinion
8. Fact
9. Opinion
10. Fact
11. Fact
12. Opinion
13. Fact
14. Opinion
15. Opinion

Water cycle

1. Accumulation - The process in which water is collected in large bodies.

2. Condensation - The process in which water vapor in air is converted into liquid water. This process is responsible for the formation of clouds.

3. Evaporation - The process in which liquid water becomes water vapor.

4. Precipitation - The process in which water falls from the clouds in the sky.

5. Runoff - Water that flows either on the ground or under it as streams, drains or rivers.

6. Transpiration - The process in which water from within the plants evaporates into the atmosphere.

Layers of the Atmosphere

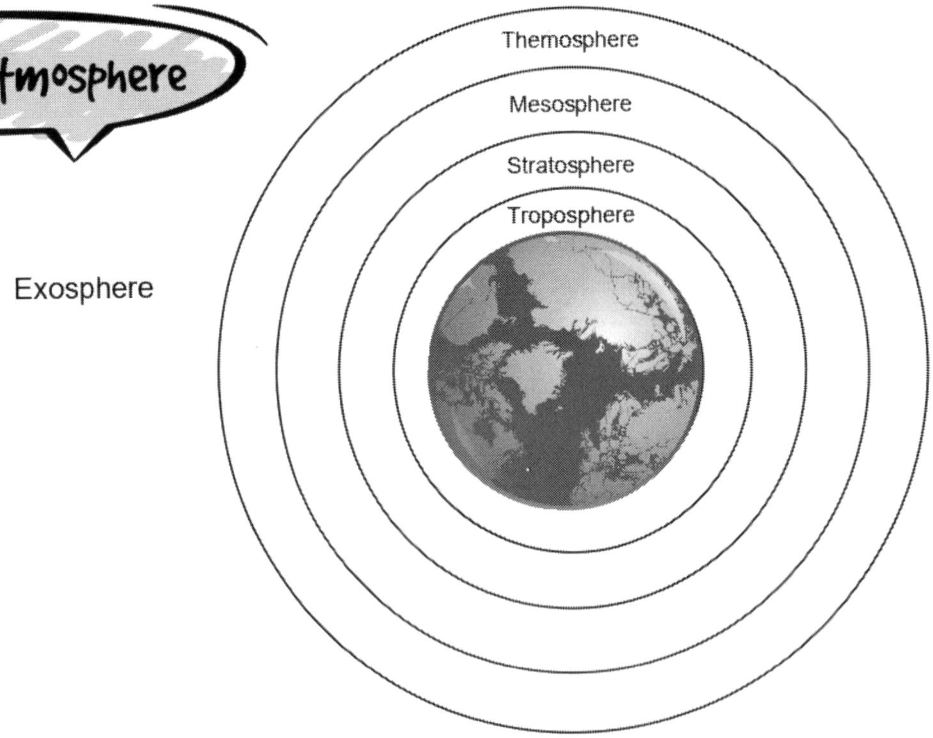

Sea Breeze Vs Land Breeze

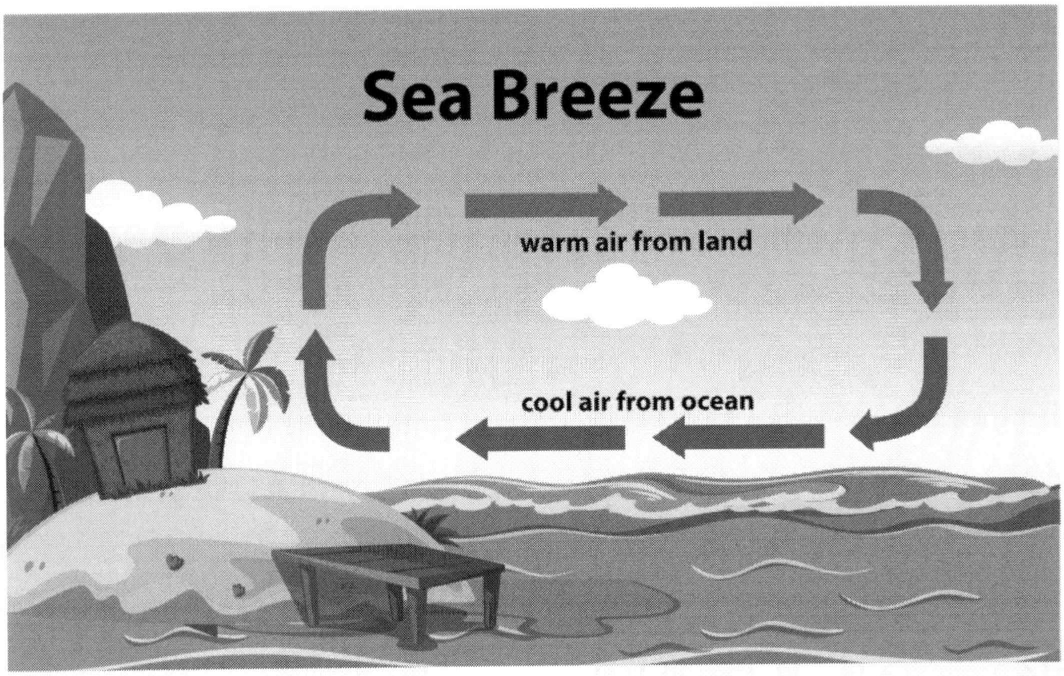

1. Sea breeze blows during the day and the land heats up faster than the sea. The air on land becomes less dense (i.e.warmer) and rises so the cooler air over the sea which is denser(cooler) flows in to take the place of the warm air, causing a sea breeze.

2. Land breeze blows during the night from land to sea and the land becomes cooler faster than the sea. The air above the sea becomes less dense (i.e. warmer) and rises. The cooler air from the land moves in to take its place.

3. Convection currents

4. Temperature differences and pressure differences between the land and water body

5. Warmer air is lighter (less dense) than colder air

6. Land breeze and sea breeze

7. Face the ocean

Weather Vocabulary

1. Tornado
2. Jet stream
3. Lightning
4. Cumulus clouds
5. Fog
6. Cirrus clouds
7. Smog
8. Thunderstorm
9. Greenhouse effect
10. Front
11. Relative humidity
12. Hurricane
13. Blizzard
14. Climate
15. Stratus clouds

ocean currents

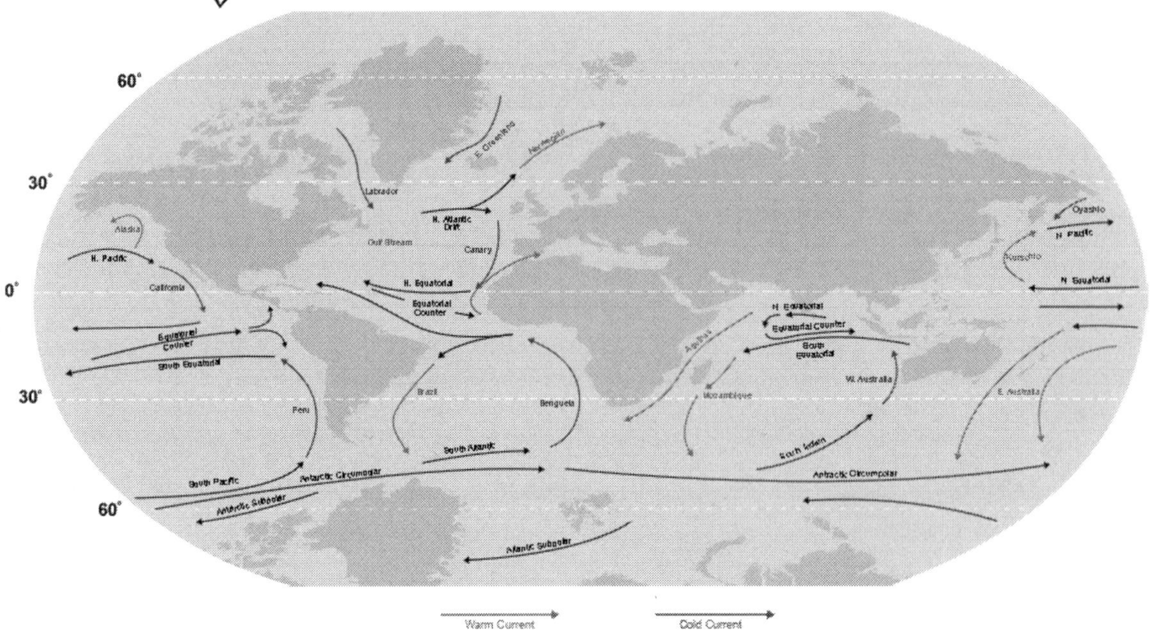

Name of the current	Characteristic temperature of the water transported by the current (cold current/warm current)
Antarctic circumpolar current	Cold current
Benguela current	Cold current
Brazil current	Warm current
California current	Cold current
Canary current	Cold current
East Australian current	Warm current
Gulf stream	Warm current
Kurushio current	Warm current
Peru current	Cold current

2. a. clockwise

b. counterclockwise

3. a. cold

b. warm

c. cooler

d. warmer

Ocean Currents and Weather

1. The equatorial currents get their names from their position with respect to the equator. The North Equatorial current is found to the north of the equator and the South Equatorial current is found to the south of the equator.

2. The Gulf Stream

3. The plate tectonics, earthquake, wind, and gravity

4. The Gulf Stream and the California current because they run closer to the United States.

5. They bring warm air into the land.

Freshwater Vs Saltwater

Freshwater	Both	Saltwater
Ponds	Has currents	Seas
Stream	Natural resource	Arctic Ocean
River	Groundwater	Southern Ocean
Lake		Indian Ocean
Glacier		Pacific Ocean
		Atlantic Ocean
		Most of Earth's water

Distribution of Water on Earth

1. a. Atmospheric water

b. Rivers and lakes

c. Groundwater

d. Glaciers and icecaps

e. Oceans

2. Oceans

3. Atmospheric water

4. Saltwater

6. Glaciers and icecaps

www.aceacademicpublishing.com

THE ONE BIG BOOK

GRADE 5

For English, Math, and Science

Ace Academic Publishing
ACHIEVING EXCELLENCE TOGETHER